Spaces of Governmentality

New Politics of Autonomy

Series Editor: Saul Newman

In recent years, we have witnessed an unprecedented emergence of new forms of radical politics—from Tahrir Square, Gezi Park and the global Occupy movement to Wikileaks and hacktivism. What is striking about such movements is their rejection of leadership structures and the absence of political demands and agendas. Instead, their originality lies in the autonomous forms of political life they engender.

The New Politics of Autonomy book series is an attempt to make sense of this new terrain of anti-political politics, and to develop an alternative conceptual and theoretical arsenal for thinking about the politics of autonomy. The series investigates political, economic and ethical questions raised by this new paradigm of autonomy. It brings together authors and researchers who are engaged, in various ways, with understanding contemporary radical political movements and who approach the theme of autonomy from different perspectives: political theory, philosophy, ethics, literature and art, psychoanalytic theory, political economy, and political history.

Titles in the Series

Spaces of Governmentality, by Martina Tazzioli

Spaces of Governmentality

Autonomous Migration and the Arab Uprisings

Martina Tazzioli

ROWMAN &
LITTLEFIELD
INTERNATIONAL
London • New York

Published by Rowman & Littlefield International, Ltd.
Unit A, Whitacre Mews, 26-34 Stannary Street, London SE11 4AB
www.rowmaninternational.com

Rowman & Littlefield International, Ltd. is an affiliate of Rowman & Littlefield
4501 Forbes Boulevard, Suite 200, Lanham, Maryland 20706, USA
With additional offices in Boulder, New York, Toronto (Canada), and London (UK)
www.rowman.com

British Library Cataloguing in Publication Information Available
A catalogue record for this book is available from the British Library

ISBN: HB 978-1-78348-103-3
ISBN: PB 978-1-78348-104-0
ISBN: EB 978-1-78348-105-7

Library of Congress Cataloging-in-Publication Data

Tazzioli, Martina.
Spaces of governmentality : autonomous migration and the Arab uprisings / Martina Tazzioli.
pages cm
Includes bibliographical references.
ISBN 978-1-78348-103-3 (cloth : alk. paper)—ISBN 978-1-78348-104-0 (pbk. : alk. paper)— ISBN 978-1-78348-105-7 (electronic)
1. Arab countries—Emigration and immigration. 2. Immigrants—Arab countries. 3. Arab Spring, 2010– 4. Social movements—Arab countries. I. Title.
JV8760.T39 2014
909'.097492708312—dc23
2014030488

™ The paper used in this publication meets the minimum requirements of American National Standard for Information Sciences Permanence of Paper for Printed Library Materials, ANSI/NISO Z39.48-1992.

Printed in the United States of America

Contents

Acknowledgements

To Bright Samson, Hamidou-Karim Soley, Hammed Parker, Jacoub and all the people at Choucha camp. For their ongoing struggle for a space to stay.

To Federica Sossi, for having shaken up this work with her radical insight. And to Glenda Garelli, for having brought everything together, without getting lost, despite the United States. To Glenda and to Federica: for Tunisia, for the book and all the books; for the countermaps drawn, the countermaps 'acted' and for our spatial insistences in the Tunisian space.

To Judith Revel, for bearing with me and for answering yet another question from me. And also because she encouraged me to bring Foucault out of the Philosophy Department.

To Sandro Mezzadra, theoretical and political yardstick of this work, between philosophy and the world. And to Nicholas De Genova, an effective master over the last four years, for his 'politics of incorrigibility'.

To Arnold Davidson, who started everything with a course on Foucault in Pisa and a question about the Intifada. And for his 'discordant practices of freedom'.

To Sanjay Seth, for re-routing my rambling digressions and my theoretical drifts from the beginning.

To William Walters, for the Foucaultian governmentality, and for our dialogues at distance.

To Claudia Aradau, for the unexpected trust in this work.

To 'my Foucaultians': Laura Cremonesi, Orazio Irrera and Daniele Lorenzini. Incomparable deskmates. For our seriousness in Materialifoucaultiani, invented four years and half ago in Paris; in part for fun, in part not.

To Alessandra Sciurba, who encouraged my first entry in the world of migrations, some years ago.

To Walid Fellah, who made the revolution that I write about.

To Monica Scafati, for making space, not borders. And to Oana Parvan, for the chicken and the egg in Tunis.

To Charles Heller and Lorenzo Pezzani, for the countermapping - our real obsession .And for Tunisia-Goldsmiths, for more or less the same thing.

To Stefania Donzelli, for the militant research, that we like so much, even when we don't know exactly what it is.

To Debora del Pistoia, for her courage to stay in Regueb, that I have not; and to Marta Bellingreri and Francesca Zampagni, in, Tunisia-Italy-and somewhere else; like me, dispersed in the world.

To the 'Londoners': Amedeo Policante, for our paper on Foucault and drugs; and Yari Lanci, for Foucault and the history of the present, an unexpected lucky shot; and, obviously, to Evelina Gambino, who 'pluralizes'.

And, finally to Valeria Stenta and Luisa Giorgetti, for our uneven mobility and for sharing philosophy since our time in Berlin and San Cristobal.

Introduction

The Arab uprisings and the practices of migration which they prompted troubled the Mediterranean order of democracy and mobility, producing spatial upheavals which reverberated also on the northern shore of the Mediterranean. This is the conclusion shared by different political perspectives and actors—liberal analysts, Left movements, policy makers and scholars: the European Union was taken by surprise by the sudden outbreak of the Arab Spring (Peters 2012), which marked the reawakening of Arab history and its '1848' (Badiou 2012); a wave of popular discontent of unexpected magnitude shook the Arab world (EC 2011); there was a tsunami of young Tunisians (Aita 2013) and Tunisians struggling and fighting against unemployment (Boubakri 2013; Carrera and den Hertog Parkin 2012). All these analyses highlight that both the Mediterranean and European spaces were shaken by the unfolding of revolutionary events and by the 'domino effect' that they triggered. However, this research, conducted across the two shores of the Mediterranean, examines the struggles for democracy in the Arab countries together with practices of migration across the Mediterranean. These two phenomena remain instead fundamentally disconnected in most analyses: unauthorized migrations are presented as a troubling factor for achieving a smooth transition to democracy. What emerges is the image of an upheaval that is definitively reabsorbed and included into the existing economic geographies.

This book investigates the political and spatial upheavals that took place across the Mediterranean, centring on the connections between practices of migration and revolutionary uprisings and looking at the reverberations in the European space. Indeed, it addresses events that are still underway, since three years after the outbreak of the Tunisian revolution the political and spatial turmoil that it sparked has still not come to an end. The linchpin which sustains this research on migrations and revolutionized spaces is the turbulence that erupted in the Mediterranean region after the outbreak of the Tunisian revolution and that was characterized by what I call a *twofold spatial upheaval*. Indeed, migrations and political uprisings during the Arab revolutions should both be considered as practices of freedom that 'disturbed' and broke the spatial and geopolitical stability of the Mediterranean. Although related and strongly connected, however, the two movements in question—migrations and the revolutionary uprisings—have their own specificity, so that it is not possible to fully conflate Tunisians' strategies of migration with the same narrative of social unrest taking place in Tunisia against the regime. For this reason, a spatial gaze on the Mediterranean turmoil needs to explore the interplay between migration movements and revolutionary uprisings, paying attention to the ways in which migrations brought upheaval to the northern shore.

The narrative of maps percolates throughout the book, highlighting the multiple and contested functions that mapping plays in the government of migration: migrants' enacted cartographies; the map as a governmental tool of knowledge; counter-mapping as a challenging analytical approach; migrants escaping the regime of representation; an ongoing widespread use of maps and visual metaphors for speaking of migration. Moreover, the topicality of maps for this work depends also on the present functioning of the technology of government over migrants' lives: as I will show later in the book, the attempts to track, spy on and map unauthorized movements, beyond the physical presence of bodies—as in the case of the traces left through biometric data—or monitoring at a distance the presences in space—as in the case of controls at sea—are what definitively characterize the present bordering mechanisms and the regimes governing migration. Rather than staying and fixing on the geopolitical line, the mechanisms of capturing migrants' movements frantically try to follow, anticipate and manage migrants' routes and the virtual or concrete traces they leave. Migrations across the Mediterranean have shaken the cartographic order of the European space, moving in unexpected ways and interrupting also the temporal progressive narrative that sustains maps: during the Arab uprisings practices of migration disjoined the temporality of the 'not-yet', through which colonial spaces have been narrated, and the usual speed of migration policies in tracing new borders. These migration practices suddenly accelerated the supposed linear temporality and spatiality of 'ordered mobility', forcing migration governmentality to frantically reassess its mechanisms of control over migrants' lives.

This book looks at migrations taking place in the aftermath of the Arab uprisings as *strategies of migration* for enacting the freedom that Tunisians won through the revolution, unsettling in this way the pace of mobility established by migration policies and by economic bilateral agreements (Mitropoulos 2007). By 'strategy' I don't mean a planned set of actions, but a practice that is undertaken for finding another space to live or for doing what 'authorized' mobile people ordinarily do, namely, move round. However, situating this research on movements of migrations in the frame of the Arab revolutions allows me to show that it is not migration per se that is a strategy of freedom. Any migration movement needs to be grasped in the light of the political, geographical and social context in which it takes place, exploring to what extent it exceeds or disrupts existing economies of mobility and techniques of bordering.

Thus, as far as the Tunisian revolution is concerned, undocumented migrants who left after the fall of Ben Ali were able to migrate because of the revolution, and their practice of migration could be seen as a way of enacting and continuing the revolutionary demands of freedom and democracy, thus disrupting the regime of conditional and selected access to mobility established by Euro-Mediterranean agreements. Moreover, the revolutionary political framework in which these practices of migration are situated requires a shift away from the border as site of exception and limit to investigating migrations in the light of broader social and economic issues.

The proliferation of mobility partnerships, economic agreements and neighbourhood policies between the European Union and the countries of the Maghreb region signals the restructuring of migration governmentality in the face of those spatial upheavals. By 'migration governmentality' I refer to the multilayered and heterogeneous set of technologies, discourses and policies concerning the production of borders and their differential functioning, and at the same time the regulation of people's movements. Taking on a Foucaultian perspective, this assemblage of political technologies is, however, seen as a contested field whose global dimension, as well as stability, is constantly challenged by conflicting actors and interests and by migrants' turbulences.

A gaze on the transformations underway in the politics of mobility and on the deep destabilization of governmental cartography makes it possible to *take in reverse* the migration regime, causing its supposed solidity to crumble. To take it in reverse means also investigating what it effectively produces beyond its texts and narratives: one of the main arguments of this book is that the migration regime works through precarization and interruptions. In fact, in the face of strategies of life that migrants try to actualize, migration policies respond by fragmenting people's journey, stranding them in certain designated places or hampering their ability to choose where to move or stay. Concurrently, it makes migrants' lives precarious, or rather, it introduces differentiated forms and degrees of economic and existential precariousness according to a variable range of mobility profiles ranging from 'economic migrants' to 'mobile person/non-migrants'. More than a hyper-governmentalization of migration, what is at stake today in the European and Mediterranean space is a sort of *desultory politics of mobility* operating intermittently and by fragmenting migrants' journeys according to a logic of *dis-charge* that works, according to different spaces and times, by taking control over migrants or by leaving them undetected.

Focusing on migrants' spatial upheavals, this book challenges government-based approaches, punctuating the way in which (some) migrant movements do not follow the governmental map and disrupt its cartography, forcing power to invent new mechanisms of capture and to arrange new narratives. A spatial perspective on undocumented migrations enables us to concentrate on the effects of unauthorized mobility and border enforcements: the turbulence (some) practices of migration engender, the interruptions they produce in the mechanisms of governmentality, and how power invents new technologies and rationales of bordering. In other words, a spatial gaze provides us with a fruitful insight on the huge productivity of borders, tracing exclusive spaces of free mobility, or zones of detention, in the face of disordered practices of mobility.

Starting from this background, I mobilize two main analytical tools: governmentality and counter-mapping. These concepts work here as operational tools for unpacking the governing of mobility and, jointly, to highlight the idea that migration as an object of knowledge and government is a contested strugglefield. But, at the same time, the use I make of these analytical tools is at the very limits of their tenability—that is, this analysis situates itself at the tipping point of these concepts and of their possible use. In fact, it is precisely by confronting issues and contexts in which these concepts 'fail' that we can envisage and craft new categories and semantic assemblages. As I illustrate in the first chapter, over the last decade the use of Foucault's analyses of governmentality has blossomed in migration studies. However, the critical and genealogical implications of Foucault's reflections on governmentality tend to get lost in these analyses and instead the focus becomes the proliferation of border controls and the existence of a global migration regime. Drawing on Foucault and pointing to the thresholds of the unacceptability of power, this work treats governmentality from a quite different angle, unfolding the constitutive dimension of struggle over mobility in the production of *spaces of governmentality*. By this expression—'spaces of governmentality'—I mean spaces that, at some point, become contested sites of movements, politics, governmental interventions and struggles: spaces that often do not coincide with geopolitical borders and that, rather, are the outcome of the emergence of a series of events, subjects and movements as a 'problem'—that is, as phenomena to govern. Therefore, these are spaces that are produced and enacted as a strugglefield of governmentality, since subjects and movements constantly challenge the governmentalization of lives and spaces that migration policies and techniques of bordering put into place. 'Spaces of governmentality' conveys the reversibility and mobility of power relations that characterizes the Foucaultian notion of governmentality, pushing us to look at migrations not merely as resistances to mechanisms of capture but, rather, as movements and subjective drives that trouble the spatial and political order of citizenship. At the same time, the concept of 'spaces of governmentality' emphasizes that the migration regime is a political technology for governing migration populations and singular bodies but is also a way for governing spaces.

The counter-mapping perspective that I undertake is part of the spatial vantage point that is the keystone of this work. It refers simultaneously to two orientations. On the one hand, counter-mapping centres on some cartographic practices that have challenged governmental maps on migrations, bringing to the fore the struggle over (in)visibility upon which migration governmentality is predicated. On the other hand, it addresses a non-cartographic engagement which consists fundamentally of dislocating the analytical posture usually adopted for looking at migrations: it gestures towards the subjects and the spaces where mechanisms of governmentality have an impact to understand their effects and how they are resisted by migrants; and secondly, it implies seeing how migrants sometimes trace *another map*, performing unexpected geographies which cannot be encoded into the cartography of government.

The use of these two concepts leads to a parenthesis on a methodology that seeks to account for the ongoing disarray of spaces and categories that percolates through the politics of mobility. Indeed, the field of migration is characterized by the frantic reassessment of policies and strategies and discourses in the face of the migration turmoil. There is a frenetic proliferation of new objects of knowledge and contested spaces of governmentality, which emerge in order to keep up with migrants' turbulence, and they are constantly defeated by strategies of migration. Moreover, the upheavals produced by migrants and processes of bordering are located in and intersect with other *spatial disarrays* concerning the government over lives: the politics of humanitarianism securitizing logics, the politics of labour, class divisions, spatial economies, and so on. For this reason a critical investigation which seeks to come to grips with the migration strugglefield and with new spaces it generates (see chapter 3) needs to equip itself with an analytical posture sufficient to the task. In this regard, a twofold simultaneous move is necessary between concepts and objects. On the one hand, as stated above, analytical tools work here as instruments to unpack and destabilize objects and categories about migration in order to grasp the mechanisms of capture they activate. On the other hand, in the place of positing an object to focus on, this study starts from different events, spaces and strugglefields tracing back how they have been framed as objects of intervention and knowledge. This entails moving sideways from bounded spaces and case-studies, unfolding them into broader political settings and articulating with processes and movements that cannot be visible on a traditional map—like, for instance, the emergence of new zones of governmentality—and that are of different natures—economic, juridical, racial. Moving sideways across bounded spaces, following the surfaces of friction where concepts at hand crumble—these are the two main methodological orientations that sustain this analysis of underway spatial upheavals. All this entails a twofold historicizing gesture: both the objects and the political-analytical categories through which objects are framed need to be questioned in their historical function and emergence. The spatial-temporal localizations turn into 'a critique that, ultimately, is both a cartography and a genealogy' (Revel 2012). This leads also to the challenging of any spatial fix and geographical blueprint: thus, the Mediterranean is not taken here as an unquestioned spatial referent; on the contrary, it becomes the object of a critical investigation that points to the transformations of its imagined and political boundaries, as well as to the effect of power-knowledge produced by the increasing centrality of the Euro-Mediterranean label grounded on a Euro-leading and of the discourse on the 'proximity' between the two shores.

The changing political context in which this analysis is situated—the Arab uprisings and their Mediterranean disseminations—requires a reconsideration of how to deal with events underway. In 1973, Michel Foucault defined himself as a journalist because of his interest in the present and philosophy itself as a practice of 'radical journalism' that tries to transform the present reality (Foucault 2001d). In this way, the diagnostic of the present is connected from the outset to a transformative work in order to overcome existing power relations. But the reference to journalism appears in many of Foucault's texts, and it is employed also in addressing events which were unfolding as he wrote, as in the case of his reporting on the Iranian uprisings in 1978–1979. In this regard, Foucault's suggestion of a history of the present represents a particularly useful insight for undertaking a close scrutiny of the *twofold spatial upheaval*, which is the subject of this work, to grasp the singularities and the discordant practices of freedom that could not fit into traditional political narratives (Foucault 1984b). To put it bluntly, this work resists framing the Tunisian uprisings according to the script of the transition to democracy and the paradigm of the Enlightenment of the Arab people. Rather, it makes us question what kinds of practices of democracy are at stake, and to what extent they resonate far beyond those spaces, unsettling in part the tenability of the European model of democracy. Similarly, Tunisian migrations are taken into account here as practices of movement that in some way push forward and enact concretely the freedom conquered in Tunisia.

Grappling with a history *in* and *of* the present means disengaging from the conceptual governmental grid on migrations, as well as from the teleological political narrative of democracy and secularism (Buck-Morss, 2009). There is no exemplary model to apply or from which to learn: this could be the general formula for summing up the attitude that a history of our present requires. Instead of looking at political practices and events and searching for what we want to find there, we should explore how current movements disrupt or drift away from established epistemic and political scripts. In this regard, Foucault's writings on the Iranian uprisings constitute an important reference for thinking of a 'diagnostic of the present' [Foucault, 1968a]. For instance, Foucault observed substantial differences between the Western revolutionary model, which 'tamed' events into a rational and progressive history, and the Iranian uprisings, in which the revolution is not limited to the fall of the dictatorship but includes another way of conceiving the political, deeply restructuring the relationship with modernity, religion and transformation. In the case of the Arab Spring and migrants' practices in the aftermath of the fall of the regimes, this analytical posture in the face of the events makes us interrogate what kind of democracy is envisaged by those upheavals.

Foucault's history of the present enables us to disconnect the analytical gaze from the validation of established teleological narratives which fix in advance the future outcomes of those events (Foucault 2001e; 2001g; 2001l).[1] Indeed, Foucault's use of history has a strategic and political function which consists in highlighting the disconnections and the discontinuities between what is happening and what we would like or we expect to take place: the concept of history which underpins a diagnostic of the present is not grounded on permanent features. Rather it aims at locating and producing discontinuity within progressive and secular narratives (Foucault 1984a). Hence, making a history *of* and *in* the present involves showing the opening of new political practices and narratives. To phrase it from a slightly different angle, migrants' spatial upheavals push us to interrogate the adequacy of the framework and the grammar through which we look at the instabilities that migrations produce (Asad 2009). Thus, the simultaneous challenging of Western political narratives for reading the Arab uprisings and of Migration Studies' Eurocentric epistemology, brings us to challenge the 'methodological Europeanism' that posits Europe as a stable counterfoil through which understanding the orientation of political struggles takes place [Garelli, Tazzioli, 2013].

POLITICAL EPISTEMOLOGY AND MILITANT RESEARCH: CHALLENGING MIGRATION KNOWLEDGE-BASED GOVERNANCE

The centrality gained by the word 'migration' in the last two decades both in the academic setting and in EU funding schemes is a clear signal of the contested nature of this multifaceted field of research, essentially formed by two entangled and co-determined notions: migration and borders. Research that aims to challenge the existence of the migration regime cannot overlook what I call the *migration hodgepodge,* meaning the entanglement of migration with other manifold political and social issues that, all together, form a sort of 'dodgy continuum' under the umbrella of (in)security: terrorism, criminality, poverty, unemployment, and marginality.

Against this background, research on migration and borders that presents itself as 'critical' cannot avoid a radical interrogation of the very meaning and possibility of critique and of its political effects and counter-effects—the implications, the epistemological coercions and the political impact of critical research beyond its intentions. To put it differently, this work enquires into the deadlocks and the difficulties of doing critical research on migration in a way that neither reiterates nor fosters the regime of truth and knowledge underpinning migration management. Research on migration is not simply enmeshed and influenced by a specific and overwhelming regime of truth that frames migration as a phenomenon to govern, but it is rather built on the same epistemological boundaries, categories and assumptions that define the epistemic community of migration studies. These issues underlie all the chapters, problematizing from time to time which analytical posture is undertaken in order not to corroborate the same order of discursivity upon which the government of migrations is predicated. Starting from these considerations, the specificity of a militant research approach consists in struggling over the knowledge production on migration in the light of migrant struggles and political experience 'on the ground' (Malo 2007; Colectivo Situaciones 2007; Counter Cartrographies Collective 2012).

Political epistemology is the name for a practical engagement with the politics and economy of knowledge on migration (Davidson 2013; Mezzadra, Ricciardi 2013): it is not a theory but, rather, a political task and an analytical posture at the same time. The posture of political epistemology consists in a twofold opening move, always starting from within. First move: it brings out the conceptual and discursive field through which the 'worldmaking' of migration governance is performed, highlighting that discourses and categories are the crystallized outcome of power-knowledge relations. Political epistemology pushes forward the task of critique, aiming at destabilizing the economy of knowledge that selects and translates some practices of movements as migrations to govern. Second move: political epistemology struggles with and opposes the juridical and political sorting machine through which practices of mobility are selected and classified, tracing exclusive spaces of mobility and protection and creating migration profiles. This second move concurrently strives for a counter-mapping gesture and the crafting of a new language for speaking of practices of movement. Indeed, we cannot avoid asking to what extent it is effectively possible to counteract or overturn the meaning and function of migration governmentality language and taxonomy, due to the complex chain of signifiers and the regime of power/knowledge in which any concept is shaped and works. Therefore, it is here that the invention of an alternative vocabulary to speak of practices of movement emerges. The hard task of transformative critique starts from the effort to not conflate practices of movement into the governmental signifier of 'migration'. However, a political epistemology on migration knowledge should not stop at this task: it should destabilize the internal coherence of the existing regime of discourse, and not be complicit with its reproducibility [Butler, 2006b]. The invention of another language for speaking of practices of movement should not be limited to an oppositional gesture which produces new categories without challenging the 'field of stabilization' and the 'style of reasoning' that underlie the rationale of migration governmentality (Davidson 2004; Hacking 2004a). It is the chain of equivalence through which some categories and discourses are linked to one another that determines the consistency of the migration regime and the naturalized assumption that migration is a phenomenon to govern. Moreover, it is fundamental to stress that political epistemology of migration actually is *not* a question of categories. This might seem an odd thing to say, especially in the wake of the above explanation, but it addresses precisely the disciplinary function of juridical and epistemic categories. To put it differently, the primary function of the proliferation of mobility profiles is not to create juridical classifications but, rather, to discipline and manage bodies and movements in spaces. This intention to go beyond categorization is thoroughly explained by Foucault in his course on Psychiatric Power: referring to psychiatric classification, he argues that nosological categories 'are not in fact employed here at all as a classification of the curability of different people'; rather they serve to define the possible utilization of individuals', the possibility to put them to work and to govern their conducts (Foucault 2006a, 128).

In fact, the 'nominalist inventiveness' of categories is associated with a specific regime of truth that traces the always changing and slippery boundaries between migrants and non-migrants. Therefore, the discursive domain cannot be taken as a self-standing reality, but rather it should be always analysed in relation with non-discursive practices that influence its limits and conditions of emergence and 'truth' (Foucault 2001a; 1980b). Instead of positing the power of categories to classify all people and divide them up into profiles, political epistemology brings to the fore the instabilities and crises that break them apart, which are generated by practices of movement that exceed clear-cut boundaries. However, critical accounts and radical theories of migration seem to be easily accommodated within the 'becoming (academic) discipline' of migration, resulting in a *discipline effect* of migrants' troubling practices. Actually, the *disciplining* of migration has been fostered by the emergence of migration studies as a regulated academic field of knowledge that posits migration as an autonomous object of research. But 'discipline' also refers to the effect of power engendered by the political 'normalization' of migrant upheavals. Indeed, firstly, migration studies tend to reiterate the methodological nationalism that underpins traditional analyses of migration, taking the space of citizenship as the vantage point against which migrants' practices are assessed. Secondly, migration studies 'tame' migration, taking it into account not as a name for heterogeneous practices that resist and escape economic and political controls over lives but as a phenomenon to regulate. Finally, migration is shaped as a self-standing research object, detached from the contested strugglefield in which practices of migration always happen. Thus, 'migration' becomes a field of inquiry and engagement for specialists—policy makers, scholars and lawyers are the figures authorized to understand the multifarious phenomenon of migration.

Conversely, a militant research approach reconnects 'migration' with the multiple strugglefields in which practices of mobility take place, looking at the spaces they produce and transform. If we shift from the paradigm of critique to a possible *dislocation of uses*, what emerges are knowledges that can be rerouted and put to work in unexpected contexts and which diverge from their original function. This is far from suggesting that migrations are narrated through neutral categories and discourses: on the contrary, the issue is to point to the leeway that always exists for disrupting a given regime of truth from the inside. The militant research approach engages in an incessant struggling *with* and within existing normative frames—asylum, rights, hosting—pushing the limits of their tenability and function. Jointly, it strives for thinking beyond these categories, exploding their meaning and the selective logic of governing mobility.

The second orientation of militant research leads us outside the boundaries of academia, raising the difficult issue of interrelations with migrant struggles: the knowledge effects, the possibility of struggling along with migrants, and finally the counter-effects of migration research—that is, the frictions, impacts and 'unwanted effects' that this research could engender. Nevertheless, it is not a matter of a univocal direction—from political engagement with migrants to the production of critical knowledge—but of a constant back-and-forth between the two. A tricky issue that characterizes a militant research approach to migrations concerns the *distance* that is at play between the political stakes of migrant struggles, on the one hand, and the field of research on the other. 'Distance' does not refer here to the gap between theory and practice; rather, it stands for the nonreciprocal nature of the relationship between undocumented migrants, who are out of the 'social contract' of citizenship, and others—including critical researchers; and it is precisely such a distance, I contend, that complicates the possibilities of tuning in to the rhythms, the goals and the languages of migrants' claims. Indeed, a common feature among the manifold effects generated by migration policies and the disciplining of migration consists in the production of distances: spatial distances, namely, the boundaries of the university but also the huge expanse of the Mediterranean that undocumented migrants must cross; social distances, with scholars and researchers as entitled with expertise and, on the other hand, the class divisions that the politics of visas produces between citizens; and finally, existential and juridical distances.

This book suggests that it is necessary to keep these impasses in the foreground, and not fully and immediately encode practices of migration into existing political claims and vocabulary, simply showing how they destabilize and trouble those frames. Secondly, it gestures towards the crafting of a knowledge practice stripped of the 'comfort of critical distance with regards to the object' (Colectivo Situaciones 2003). From this standpoint the question is not so much to narrow or bridge these distances and gaps, but rather to put them to work and make visible how (some) migrations play in a discordant way in relation to dominant maps and languages. It is only by seriously taking into account this produced but existing distance that it becomes possible to work through it. Indeed, militant research doesn't aim to fashion the law on migrant struggles. On the contrary, it involves making those troubling mobilities resonate within the domain of knowledge production and, starting within the epistemic community of knowledge on migration, trying to unsettle the legitimacy of this domain by following constituent movements that open spaces of struggle beyond it. For this reason I prefer rephrasing the 'within and against' workerist formula[2] in terms of 'within and beyond': struggling within the existing epistemic and political space to build and open alternative practices for thinking and dealing with people's movements.

'Militant research' refers to a theoretical and political engagement in the contested field of migration that, aware of the crucial role of knowledge in tracing new borders and inventing apparatuses of capture, gestures towards an interference between involvement in struggles and the space of knowledge production. It strives to hijack and reroute knowledge by putting it to work in contexts other than those expected by the governmental and academic economies of knowledge.

NOTES

1. What Foucault highlights in the events of the Iranian revolution is the refusal of the Iranian people to play the game of politics, as it was traditionally structured, and to accept that regime of truth: *'le peuple Iranien fait le hérisson. Sa volonté politique est de ne donner pas prise à la politique'*. (Foucault, 2001b, 702).

2. According to the Italian workerist tradition of the 1960s and 1970s, the struggle against the capitalism system of production and exploitation should necessarily start and take place within the space and the conditions of capitalism itself, trying to reverse its functioning. From a more theoretical point of view, workers' labour force is conceived as playing simultaneously within and against capital, as functional to the reproduction of the capitalistic system and, at the same time, exceeding the established relation between labour and capital.

ONE

Border Interruptions

Working with Foucault between Migrant Upheavals and Spaces of Governmentality

PREAMBLE

September 20, 2011: The detention centre of Lampedusa—the southern outpost of Europe—was burnt by Tunisian migrants refusing to be deported back to Tunisia. Lampedusa's residents responded by sparking off a guerrilla campaign against migrants, and the Italian government moved Tunisian migrants into floating prisons, while the island was declared a 'non-safe harbour'. These striking events were part of what, on the European side, was designated as the 'Arab Spring' or the 'North Africa emergency'. After the outbreak of the Tunisian revolution, thousands of Tunisian citizens left Tunisia by boat towards Europe and almost twenty-seven thousand reached the Italian coasts. Meanwhile, due to the Libyan conflict, nearly one million people crossed the border between Libya and Tunisia: many of them have since been stranded as asylum seekers in Choucha refugee camp at the Ras-Jadir post along the Tunisian frontier with Libya, while around twenty-five thousand arrived in Italy claiming asylum.[1]

This preamble could have started with another date, April 5, 2011, when the Italian government signed a bilateral agreement with Tunisia to manage arrivals of Tunisian migrants, separating those who arrived by that date and who could obtain a temporary humanitarian permit, from the others, who became 'clandestine'. Or also on February 12, 2011, when Italy declared a state of emergency on the national territory, and on April 7, 2011, when, bizarrely, the decree was extended to the 'territory of

North Africa'. But, despite all these possible landmark dates, I do not focus on the events created by governmental statements, juridical decrees or exceptional measures. Rather, I trace a map of those spatial upheavals starting from and following the turbulent patters of migration produced in the Mediterranean space at the time of the Arab uprisings. The map that I will try to trace in the next chapters aims at providing a cartography of the Mediterranean instabilities as an alternative to the map produced by governmental agencies; a map of snapshots that in some way retraces, fosters and pursues the spatial and political transformations triggered by those movements. However, in order to trace 'another map' of the Mediterranean focusing on the spatial upheavals produced by the Arab uprisings and migrants' practices, we should also look at how power counteracted these eruptions. Migrants' practices do not take place in a vacuum, hence bordering and containment responses cannot be overlooked. September 23, 2011: all migrants detained in the centre of Lampedusa were moved onto boat prisons located in the sea off Palermo. A few days later, all of them were deported to Tunisia.

This chapter is formed by two main sections corresponding to the ways in which I deploy Foucault's thought in this analysis: a critical account of migration governmentality and an investigation of the politics of truth at stake in the government of migrations. In the first section I deal with the notion of migration regime, challenging its supposed stability and all-encompassing grasp: instead of starting from the existence of a border regime,[2] I will reverse the gaze, looking at migration governmentality as a complex and temporary assemblage of strategies for taming and channelling migration turbulence (Papastergiadis 2000). Starting with the idea that migration governmentality rests on a regime of truth, in the second section of the chapter I engage in an in-depth analysis of the discourse of truth that is at stake in the government of would-be refugees.

Migrations and borders are topics that Foucault never tackled; although, especially concerning borders, many scholars draw attention to passages in *Security, Territory, Population* where Foucault stresses how the management of the circulation of goods and people plays a central role in the functioning of liberal societies (Elden 2007; Fassin 2011; Mezzadra and Neilson 2013; Walters 2011a)]. The lectures at the College de France on *The Birth of Biopolitics* where the figure of the migrant is framed in terms of human capital is assumed both in Foucaultian scholarship and in migration studies as an important reference for analysing the rationale of migration governmentality (Cotoi 2011; Nail 2013; Read 2009). However, the way in which I mobilize certain Foucaultian methods and categories neither interrogates what Foucault said on the topic of borders and mobility nor takes the Foucaultian toolbox as a univocal analytical grid over-coding present events (Deleuze and Guattari 2004; Hindess 1997). Indeed, this work essentially grounds itself in a history of the present(s);

and this is the reason why one of the preliminary questions is to problematize what it means to work with Foucault today, from within the cartography and the tempos in which we live. Nevertheless, the point is not to test Foucault in the light of contemporary issues and events or to demonstrate that his toolbox is more usable than others. Rather, what is in need of clarification is the very notion of the 'use' of Foucault that has gained ground in the last decade (Barry, Osborne and Rose 1996; Burchell, Gordon and Miller 1991; Elden and Crampton 2007; Potte-Bonneville, 2004). Thus, before delving into the specific topic of this research, I linger both on the idea and on the practical engagement of using the Foucaultian toolbox.

THERE IS NO GRID FOR THE PRESENT(S): 'MAKE THE MAP, NOT THE TRACING'. INTERMEZZO ON THE 'USE'

In the last two decades, a growing literature has mobilized Foucault's work, making it travel in other spaces and contexts or deploying it to analyse social phenomena not discussed by Foucault himself (Bygrave and Morton 2008; Dillon and Neal 2008; Inda 2005; Jones and Porter 1994; Rose 1999; Stoler 1996; Young, 2001). However, confronted with that, today what needs to be interrogated concerning the use of Foucault is the pertinence of a Foucaultian grid to travel across domains and spaces. Actually, the approach of a history of the present should refuse to instantiate a main signifier of intelligibility, turning instead to the tracing of *a cartography in-the-making*, a map produced from within the present and attentive to the discontinuities at play. No single grid touches the complexity of economic processes, regimes of truth and mechanisms of subjectivation through which the current geographies of power work. On the contrary, as Foucault himself contends, the critical force of an analysis relies precisely on the refusal to superimpose a unifying grid or a principle of intelligibility.

If the choice of the Foucaultian tools put to work cannot be made other than case by case, a general orientation can however be suggested: especially in the analyses of migrations, what makes the difference between one critical reflection and another is the kind of gaze that is exercised, or better, its orientation. For instance, some scholars tend to give prominence to the ordinary working of border mechanisms and their displacement before and after geopolitical boundaries—through the implementation of the visa system or techniques of control-at-a-distance—while others stress the violence of/at the borders and the production of a border spectacle. Both these perspectives bring out relevant aspects of the functioning of the migration regime: what makes the difference between these two approaches is, I suggest, the *gaze's orientation* that in one case focuses on the differential working of the boundaries, while in the other it centres on the general mechanisms of control. In fact, any analytical gaze builds upon a certain regime of (in)visiblity, partitioning between zones of invisibility and visibility that determine which subjects are in the focus of the analysis, who is left off the map and what mechanisms of power are made visible. In the field of migrations, the thresholds of visibility and of representation determine which practices are considered 'political' and which are instead unheard, 'noisy' or 'pre-political'. In fact, as I will show later about counter-mapping, political and visual representation go together, and consequently maps are crucial subtexts for unfolding the specific regime of visibility of any politics of representation.

Working with Foucault should avoid falling into what William Walters called the risk of 'applicationism' (Walters 2012). Without translating Foucault's thought into a multifunctional task, such an analytical posture consists in detecting and troubling the thresholds of perceptibility and acceptability of power,[3] assuming as a vantage point the limits of power and its margin (Foucault 1980a) to see how the 'inside' is produced and sustained by processes of exclusion and by resistances (Foucault 1995; 2006b). Getting closer to the issue of migration, I contend that working with Foucault at the borders firstly materializes into analyses that pay attention less to borders as geopolitical lines than to the practices of bordering (Balibar 2004; Rumford 2006; Walters 2006). These latter refer to the array of heterogeneous regulative techniques and knowledges that determine the 'disposition of men and things in space' and that respond to a double function: selecting and containing mobility (Foucault 2009).

MIGRATION GOVERNMENTALITY BEYOND POWER'S TEXT-NARRATIVES?

Secondly, working with Foucault entails taking a step back from the narrative of governmentality and its supposed solidity and analysing it in relation to other discursive and non-discursive practices. Instead of assuming a functionalist approach, which assesses the success of governmental programs on the basis of the conformity between texts and reality, Foucault focuses on political technologies, suggesting a shift to the strategic reinvestments and to the unexpected effects of power's mechanisms. This entails that gaps and 'failures' in the mechanisms of governmentality are not due to occasional resistances or as side externalities; in fact, if we assess the analytics of government by its effectiveness, this would coincide with falling into the trap of trusting in a progressive rationalization of technologies (Lemke 2013; Murray Li 2007; Patton 1998). On the contrary, working with Foucault requires us to take resistances, escapes and practices of migrations as a constitutive part of governmentality: the clashes and the frictions between governmental programs and their realization on the ground signal that governmental strategies are situated within, and respond to, strategies of migration (Bojadžijev and Karakayali 2010; Mitropoulos 2007). However, the caution we take towards the texts and the narratives produced by migration agencies and states about border controls and the supposedly overwhelming visibility of monitoring techniques should not cause us to overlook the impact and the material effects of discourse in migration governmentality. On the contrary, to challenge the language of failures and gaps in talking about the migration regime means not assessing the 'effects of reality' of texts and discourses through their actualization in non-discursive practices of government. In this regard, Foucault's reflections on the specific dimension of discourse turn out to be quite apt—on the one hand, in not taking texts as the source of migration governmentality and, on the other, in grasping the concrete effects of governmentality that they nevertheless engender. This is particularly important in the context of migration and borders, due to the proliferation of documents, reports and discourses on migrants and migration, and, simultaneously, due to the ascertainment of a tangible discrepancy between those narratives and the working mechanisms of capture, selection and control. As Foucault states, to tackle such a discrepancy one should explore neither the 'real' as the expression and actualization of texts nor discourses as instruments for knowing the 'real'. Rather, this involves interrogating 'what is the "real" that discourse consists in' (Foucault, 2014, 227), namely, the peculiar materiality and effectiveness that texts have beyond the discursive dimension.

Before providing a definition of governmentality and of the way it is used in relation to migration within this book, it is important to stress that by introducing this notion in 1978, Foucault has definitively overcome the image of power relations in terms of rapports of force. Or better, the notion of governmentality opens up a strugglefield where power relations (as more than the oppositional terms 'power/resistance') are conceived as dis-symmetries that have the possibility to govern the lives of the others and to structure one's own field of action and movement. From this perspective, subjectivation does not refer merely to the effects of practices of resistance but, rather, to processes that are constitutive of the very existence of a certain field of governmentality. As Foucault argues, what defines a relation of government is not a set of forces but the governability of certain subjects and the threshold of the intolerable.

In this regard, migration represents a particularly useful strugglefield for thinking about governmentality and to see how the instability of power relations and the dis-symmetries at play are effectively enacted. In fact, if we consider that governmentality concerns, at the same time, both individual conducts and that of populations, and that, on the other hand, the strugglefield around the governing of lives is precisely at the core of governmentality, then we can see that migration is one of the contested issues in which all these elements acquire a particular salience. Indeed, the migration regime is more than a government of movements; it is a technology for producing subjects—migrant-subjects—and for governing populations and people's lives. Meanwhile, borders are not simply technologies for controlling and hampering movements, but are also the response to troubling mobilities that disrupt the order and the space of citizenship. Therefore, migration governmentality is in some way the exemplary site of the struggle around the governability of bodies, movements and subjects. Secondly, beyond the production and government of subjectivities, the migration regime brings to the fore an issue concerning the *spatial ontology* of governmentality. Indeed, the contested field of migration, shaped as a phenomenon to govern, is formed by a heterogeneous production of spaces and works through the articulation of different spatial regimes. In other words, gazing upon migration governmentality allows us to get to grips with the spatial assemblages that characterize a regime of governmentality, making clear that something like a homogeneous international or global space does not exist (Vrasti 2013). Finally, the multiple ways through which the 'migration turbulence' tends to be channelled and governed illuminate the different power technologies at stake, challenging the image of a single governmental rationale in which sovereign and disciplinary powers remain backstage. In fact, the contested strugglefield of migration and the intermingling of sovereign, administrative, governmental and disciplinary techniques of government make us reframe the very Foucaultian notion of governmentality in a more nuanced way, taking into account the coexistence of apparently discordant mechanisms of containment and subjectivation. Among IR scholars, Foucault's notion of governmentality has been largely used since the early 1990s for reformulating power relations at a global level and beyond institutional sites. Foucault's theory of governmentality has also been criticized for the abstract and totalizing character that governmentality as a liberal technology of government seems to postulate, and for its Euro-centric model that cannot account for uneven and hybrid regimes of power (Chandler 2009; Joseph 2010; Selby 2010). However, as Wanda Vrasti contends, if 'what is needed is a much more empirically rigorous investigation of the conditions and structures under which the management of populations and states becomes effective' (Vrasti 2013, 55), then the question is not to distinguish liberal and non-liberal technologies of government; rather, it is important to 'recognise the links between liberalism, global capitalism and imperialism' by focusing on the polymorphous ways in which (economic) liberalism works beyond liberal democracies, through a different combination of disciplinary, governmental and sovereign power (Vrasti 2013, 56). Against this background, one of the main uses of the concept of governmentality in this book lies in its potential to challenge any unitary notion of space that underpins global and international studies. Indeed, my argument is that the supposedly abstract nature of governmentality is related to an analytical level—governmentality as an analytics of power relations (Kiersey 2009)—while from an empirical and political point of view, it is scrutinized by Foucault in relation to specific historical and spatial contexts. Talking about migration in Europe enables us to bring into the field of governmentality coexisting heterogeneous techniques of government that do not belong to 'other spaces' but shed light on the postcolonial dimension of Europe (Selmeczi 2009).

DID FOUCAULT DECOLONIZE POLITICS?

This analysis does not concern migration per se but migrants' spatial upheavals in connections with the Arab uprisings and their reverberations in the Mediterranean and in the European spaces. This brings up the widely debated postcolonial question about Foucault's work: why utilize Foucault in this way, considering that he was largely criticized for not taking into account the colonial question and, more broadly, for not complicating his genealogies with non-Western modernities? It is indisputable that Foucault did not take into account non-European spaces and did not complicate the genealogy on the modern Western subject with other genealogies. However, his philosophical approach is particularly trenchant in order to not reiterate existing political cartographies to encode practices of freedom that resist being translated into the boundaries of democracy and citizenship as conceived by Western political thought.

Foucault has never directly coped with issues like colonialism, migration, democracy and the crisis of the nation-state, but in some way it is precisely *because* he did not conceive a theory of democracy or a theory of citizenship that his analysis makes visible what exceeds and cannot fit into existing political coordinates. It is not because one names and addresses the issues that relate to the blurry domain of the postcolonial that the Eurocentric posture is automatically challenged. In fact, in most analyses the scripts of citizenship and democracy are assumed for reading underway social and political phenomena by simply 'stretching' the borders and rearranging the codes of the space of citizenship and democracy, without really questioning their political desirability. To put it differently, Foucault allows us to disengage from the political conceptual field through which heterogeneous practices are usually related to each other—freedom and democracy, immigration and integration, cosmopolitanism and differences, representation and politics—by reformulating politics in terms of power relations and resistances. Both the assumption of common political referents—like the state or democracy (Foucault 2009: 2010)—and the designation of a pure political space are excluded from the beginning. Or better, Foucault rethinks power itself, detaching it as much as possible from a supposed 'pure' space of the political, framing it in terms of productivity, force and government over life, namely, as economy (Macherey 2013). After all, Foucault's incisive critique of the idea of a rationality[4] that works as the yardstick of practices and governmental technologies (Foucault 1994d; 1980b) helps us avoid reading the Arab revolutions in terms of the route to democracy or according to the script of secularization. While political and philosophical Western thought framed and tamed the upheavals through a historical progressive telos, the aim of a critical gaze on the Arab uprisings consists in bringing out what escapes, exceeds or does not fit into the existing political narrative and the thread of history. This analytical-historical posture is what fundamentally characterizes a history of the present (Foucault 1984b; Revel 2013a): the reference to events or singularities that burst out, wresting subjects from themselves and from the present where we are, indicates that an analysis of the present entails a constant dislocation from the spatial coordinates that define the (political) reality to which we belong.

In a similar vein, the migrant struggles that I take into account here are characterized by the staging of a disengagement from the 'civic pact' (Azouray 2008) and of a refusal in the face of 'knowledges over lives', which produce subjects through specific technologies of individualization and dispositives of spatialization. They convey an essential reluctance to fit into 'sovereign and representational dispositions' (Mitropoulos, 2007) in which strategies of migration should be read against the backdrop of governmentality.

GOVERNMENTALITY: UNPACKING THE CONCEPTUAL LINCHPIN OF MIGRATION STUDIES

In the last two decades migration has become more and more a self-standing disciplinary domain, while in the political debate it emerges always in relation to other 'social questions' (security, welfare, social integration, terrorism, etc.), working as an underlying 'hidden' subtext and as an enchaining signifier of all these other issues of government. Migration comes out as a political issue situated *à la marge* of the salient themes and then re-emerging instead when an emergency is declared. Starting from these premises, Foucault's definition of 'dispositive', then, has been reframed in critical migration analysis in terms of 'regime' (Hess 2012; Karakayali and Tsianos 2010). Indeed, a dispositif is 'a thoroughly heterogeneous ensemble consisting of discourses, institutions, architectural forms, regulatory decisions, laws, administrative measures, scientific statements . . . a formation which has as its major function at a given historical moment that of responding to an urgent need. The apparatus thus has a dominant strategic function' (Foucault 1980b, 195–96). In this regard, Foucault suggests that 'What is needed is a new economy of power relations . . . taking the forms of resistance against different forms of power as a starting point and using this resistance as a chemical catalyst so as to bring to light power relations'; concluding that instead of studying power from the point of view of its internal rationality, we should be 'analysing power relations through the antagonism of strategies' (Foucault 1982, 779). It is precisely by hinging on destituent practices opposed to the state-citizen game that one might challenge the supposed solidity of governmentality, highlighting the ongoing instabilities upon which it is constituted.

The governmental paradigm has become the encoding frame in which to think about migrations, and in turn, migration is crafted as an object of government, around which a univocal 'politics of translation' is at stake (Mezzadra 2007; 2011d): migration policies and critical analysis operate an immediate and full translation of some strategies of mobility into migrations to be managed, and 'migration' is assumed as the natural genitive attribution of governing. Within this frame, the Foucaultian notion of governmentality covers a wide range of perspectives, from the policy-oriented works and problem-solving approaches to critical knowledge and new reflections on citizenship and sovereignty (Aradau and Van Munster 2007; Bigo 2002, Fassin 2011b; Lippert 1999; Ong, 2006; Rudnyckyj 2004; Haar and Walters 2005; Walters 2011a; Inda, 2005). Governmentality is definitely the most widely used Foucaultian tool in the domain of migrations and borders. What all these different approaches to migration that make use of governmentality share is that they pay particular attention to the multiplicity of actors forming a supposedly coherent migration regime and to the multilayered structure of government together with its flexibility and capacity to re-adapt its strategies. In fact, the grid of governmentality makes possible a shift away from a sovereign-centred reading of migration—in which the nation-state as the main holder of the monopoly of migration controls is inscribed into a broader 'methodological nationalism' (De Genova 2005, 2010b; Giddens 1973, 1975; Martins 1974) and 'state ontology' (Truong and Gasper 2011). Moreover, what comes to the forefront through the diagnostic tool of governmentality is the diffraction of borders and the non-coincidence with geopolitical frontiers, produced mainly by the politics of externalization (Boswell 2003) and through the invention of embodied borders or technological borders working at a distance (Amoore 2013). Finally, governmentality points to the unceasing redefinition of borders and the complexity of a migration regime in the making as a provisional outcome of practices of migration and techniques of capture (bordering). Thus, while the term 'migration governance' (Betts 2011; Cassarino and Lavenex 2012) encompasses the intermeshing of different and sometimes conflicting practices into a horizontal and compact image of the migration regime, governmentality alerts us to its friability, presenting its supposed consistency as the provisional outcome of conflicting discourses, strategies and border struggles (Lemke 2012). If we want to trace a cartography that shows reinforcements and discontinuities between practices of movement and techniques of bordering, we must keep together the two levels of governmentality as framed by Foucault. In fact, governmentality refers both to a historically determined configuration of power and to a diagnostic tool for setting fields of problematization and framing powers and resistances into a *strugglefield* (Foucault 1996a, 2009). In fact, migration controls and bordering techniques are constantly forced to reinvent themselves, reassessing strategies and discourses in the face of practices of migration that do not work simply by counteracting a border regime that is already there, but rather (sometimes) producing 'voids'—for instance, crises in the functioning of partitioning categories and profiles of mobility—that cannot be filled by subject-positions and existing geometries of representation; or they make power respond 'within the asymmetry' between the mechanisms of capture and migrants' turmoil in spaces (Revel 2008a, 2011; Sossi 2012).[5]

At this stage it is important to closely interrogate Foucault's definition of 'government', considering it both as a reformulation of power relations and in its specificity. Indeed, contrary to appearances, government is not simply another name for power relation, at least if this latter is conceived in terms of rapports of force. This becomes clear in the conversation of 1980 *Débat sur la vérité*, where Foucault clarifies that the notion of 'government' fundamentally designates an imbalance: 'power relations are not only pure and mere relations of force . . . they are organized through an imbalance that gives to someone the possibility to act upon the others, and these latter have not the same possibility of acting as the former. . . . This imbalance is what I would call "government"' (Foucault 2013b, 117). While governmentality studies tend to deploy the concept of government on a surface where all subjects are horizontally situated in relation to one another, Foucault's definition allows us to see the gradients and the grooves upon which a space of governmentality—and consequently also forms of domination—could be built.

DISCORDANT PRACTICES OF FREEDOM

Tunisian migrants' spatial upheavals push us to reflect on the notion of government and on its limits for grasping the transformations engendered by (some) migration movements. Such a critical account is in turn related, I contend, to the issue of the codes and language through which migrant struggles are framed. I take here government in the Foucaultian sense, namely, as an action upon the possible actions of others, and therefore as the possibility for conducting oneself differently. When we look at migrant struggles and resistances, we might rightly assume that migrant strategies of resistance actualize the refusal of having one's own life and mobility governed by border controls and migration policies—and thus they always convey a fundamental will to not to be governed in such a way. But the paradigm of government doesn't enable us to thoroughly capture what (some) migration movements put into place and generate by disrupting actual political cartographies. To put it differently, migrant struggles are always strategic movements resisting a certain configuration of power relations within a given of governmentality. Moreover, if migrants' production of spaces means the enactment of practices of subjectivation and forms of life, it follows that what drives these desires is the possibility of putting into place a different way of acting and relating to oneself and others. However, framing all this in terms of 'conduct' conveys two main thorny issues. First, the couplet government-conduct and the correlate domain of action implicated in it—encapsulated in the aim of conducing oneself differently—restores a model of subjectivity characterized by the will and the ability to exercise self-mastery, universalizing in this way a peculiar model of subject as a matrix of relations with oneself and with others. Therefore, is it (only) in terms of conduct that we can designate the practices of discordant freedom enacted by some practices of migration? A first attempt to answer the question consists in trying to detach practices of freedom from the register of conduct to see if there are processes of subjectivation that do not necessarily pivot on self-mastery and autonomy. Second, it involves thinking about practices of resistance and subjectivation other than ones that start from an exclusive focus on and relation to oneself and then possibly become collective, and instead about those that are immediately conceived on a collective dimension. For this reason, instead of taking both the idea of autonomous migration and the couplet 'government-conduct' as unquestioned analytics, we must consider the slippery condition in which undocumented migrants often find themselves—forcing them to invent strategies for readapting the will to move—which actually complicates any will to conduct oneself. For instance, unauthorized migration entails first of all a change in status, namely, the fact of becoming defined as a subject illegally in movement. Tunisian migrations represented at the same time a departure from the civic norms of the state, the designation by migration policies as 'illegal' and the attempt to get rid of it by reinventing and subjectivizing themselves otherwise. This point brings up a broader issue beyond migration that concerns what can be called the subject's split condition: on the one hand, this refers to the multiple predicates and determinations that we are simultaneously shaped by and 'attached' to—like the fact of being at the same time part of a potential labour force, unemployed, migrants, women, and so on—and whose combination makes it difficult to identify a distinct 'self' that one can relate to; on the other, it stresses that subjects are constantly 'torn out' from themselves due to the many transformations of juridical status and social positions they experience, as well as the conflicting sites of subjectivation and subjection in which they are simultaneously involved (Mezzadra 2014).

The hypothesis that I propose here is that we should try to look at what—along with the actualization of different ways of conducting oneself—is produced and put into place by migrants that cannot be easily and fully captured in the relational schema of government. In fact, if in the case of the Tunisian uprisings the will and the imaginary of a concrete political alternative—well beyond the institutional dimension—was clearly at stake, it should be asked whether in Tunisian migrations the same striving for freedom was enacted on other bases. Confronted with Tunisian migrants' claim, 'I want to move from Lampedusa and arrive in France', we should ask whether it is the 'will' that drives those practices or if, for instance, it is more a question of being part of and living in a certain space—that then immediately becomes the common imagined space for all those who left Tunisia. Thus, from this standpoint the question becomes one of understanding the discordant movements and practices that migrants engage in; what is produced or transformed that cannot be mapped or explained through the existing relational frames. For instance, what does production of spaces precisely mean and what kind of production is at stake? Tunisian migrations acted by emptying, destituting and de-codifying existing political frames and languages, enacting spaces of movements and existence that pass under the threshold of existing maps. The opening of new spaces of movement and the practices of staying in those places could be compared to the enactment of new forms of relations and to the emergence of strategies of life. And their troubling element consists in the difficulty of encoding or fully reabsorbing them into existing social relational schemas and mechanisms of juridification of existences—namely, the mechanisms that fix people's identity, status and location according to juridical categories and binary divisions.[6]

Hinging on the above considerations, 'migration government' should not be used as a trans-historical catchword for addressing different contexts, in order not to fall into the trap of 'presentism', namely, to make an analysis that erases the historical thickness of concepts and practices (Ong 2006).[7] The emergence of migration government as an object of the discursive regime and as an overarching concept traversing disciplinary domains is a quite recent phenomenon, dating back to the 1950s. It was only through a series of juridical and political steps that something like a global migration regime was shaped, reaching its current juridical frame and political spread only in the early 1990s (Geiger and Pecoud 2010; Ghosh 2007, 2012)[8] even though a binding international legal regime does not actually exist, yet. Only between 2003 and 2005 did migration definitively become a central issue on the global policy agenda. But the emergence of a global regime of border technologies through the instantiation of common standards and identification systems and controls traces back to the last two decades of the nineteenth century. As Adam McKeown puts it: 'The global system of migrant identification and control is not inherent to the existence of an international system. It was a fairly late development. . . . Most of the basic principle of border control and techniques for identifying personal status were developed from the 1880s to 1910s' (McKeown 2008, 2–3). And the migration regime was not even the outcome of a structural necessity inherent to the international system; rather it emerged 'out of attempts to exclude people from that system' (McKeown 2008, 2–3; see also Geiger 2013).

The quite recent emergence of migration as a stable and coherent object of government suggests that the migration regime itself can be framed through the image of a *strugglefield*, in order to stress its contingency and its contested nature.[9] Conceiving of it as a strugglefield means, paraphrasing Foucault, that migration policies and migrants' practices 'each constitutes for the other a kind of permanent limit, a point of possible reversal' (Foucault, 1982, 279). In addition to that, talking about a migration strugglefield for designating the complex articulation between different layers and mechanisms of power-resistance could be seen as a reformulation of Marx's notion of capital as a social relation: the migrant condition is always slippery, since it is involved in a field of changing relations of power. This refers simultaneously to two aspects: first, it means that being a migrant is not a natural condition but it is rather the outcome of political technologies, geopolitical asymmetries and class relations which come to shape the migrant condition as a social total fact (Sayad 2004); and second, it stresses the fundamental productivity of the migration regime in shaping and fixing identities. Nevertheless, the image of the strugglefield underlines the strategic and conflicting dimension in which the migration game is played. Consequently, it is not a 'field' perfectly distinct from others: on the contrary, as a set of social and power relations it influences and articulates with what is supposed to be outside of it.

However, in the field of migration studies, the entry of governmentality as a grid came to foster the logic and discourse of government, envisaging and speaking of a space of managed mobility: the stress on the heterogeneity of actors has, on the one hand, flattened the conflicting dimension underlying the two main questions, 'who governs?' and 'at what price?', and, on the other, enhanced the idea of a smooth functioning of the mechanisms of control grounded on the coordination of multiple agencies. In this way, the productive and troubling force of resistances and migration practices within the architecture of migration government is eclipsed; but in fact, practices of migration and resistance at the border trouble the tenability of existing geographies of power, producing considerable re-arrangements in the 'migration apparatus' (Feldman 2011). By fully assuming the vocabulary of migration policies and leaving unquestioned the postulate of migration as a phenomenon to govern, the governmentality grid works as a 'reinforcement operator' fostering the scaling up and multiplication of *mobility profiles* (illegal migrant, economic migrant, asylum seeker, bogus refugee, refugee, high-skilled migrant, etc.). Against this backdrop, this book tries to reverse the paradigm of government and its supposed consistency: instead of assuming the nexus mobility-government as a point of departure, I retrace the frantic, haphazard and piecemeal nature of attempts to capture and manage migration movements.

From an analytical point of view, I focus on the border interruptions at stake in migration governmentality across the Mediterranean. *Border interruptions* refer to the disruption of the functioning of borders and bordering mechanisms that subjects and movements succeed in producing at a particular moment in time. At the same time, they relate to the spatial fragmentation of people's patterns and the temporal suspensions that borders produce—namely, the effects of borders. In fact, it is precisely this ambivalent and contested dimension of borders that the notion of strugglefield encapsulates. Interruptions *of* borders, and interruptions *through* borders: the relational and conflicting nature of the border working is precisely the marker of its reversibility and what structures the field of governmentality as a disequilibrium of mobile power relations. The notion of interruption is conceived here as a disruptive movement that does not resolve into the breaking moment but that engenders spatial transformations and processes of subjectivation. The interruption could entail the misfire of a certain mechanism of power and capture, or the impact that such a mechanism has on lives. But, as interruption *of* borders, the interruption conveys also an ethical and political gesture that resists or exceeds the existing cartography of power relations: it is precisely what spills over a given regime of truth, introducing elements and practices that cannot be fully and immediately translated or codified by it. From a theoretical point of view, such a focus on border interruptions is also a way for resisting a spatial analysis that flattens the migration strugglefield to an issue of movements and cartographies (Derrida, 2005).[10]

Foucault's analytics of power certainly brings into focus the essentially productive dimension of power's mechanisms, without however denying or overshadowing the repressive local functions at work in a broader economy of power. The first Lectures on *Security, Territory, Population,* about the relationship between space, circulation and mechanisms of security, add an important contribution: not only do the politics of migration incessantly produce and reproduce borders, but borders themselves are producers of differences, as they complicate and fragment the geometry of spaces. However, beyond evading some mechanisms of capture, at times practices of migrations temporarily crack the governmental matrix and the relationship between governors and governed, making tangible the unbearable nature of power over lives. These *border interruptions* at times take place through the practice of an illegalized mobility which does not respect the temporal pace of the authorized politics of mobility (Garelli 2013), disrupting the very logic of belonging and making unworkable the partitioning between different migration profiles. All this was particularly evident with the arrival of Tunisian migrants in the European space in 2011 and 2012: they didn't claim for asylum or protection; rather, they staged 'groundless' practices of movement, an 'unentitled' freedom exceeding the channels of the expected forms of mobility. 'We don't want to stay in Italy; we want to move, going to Europe, to France. . . . We don't need anything from the Italian government, only to be released from this tiny island and a paper to move'.[11] The majority of them did not intend to settle in Italy but to move onto France or Belgium. Indeed, the Italian temporary permit[12] that some of them obtained shifted away from both the narrative of integration and from the logic of rights, and Tunisian migrants used it as a legal tool for crossing the Italian border and moving to northern Europe and to France. Moreover, despite Tunisian migrants also being aware of the hardening of Italian laws on deportation, they were driven by an unyielding desire, irrespective of the difficulty of finding a job in Europe—to leave. At the same time they didn't see Europe as an idyllic space for human rights; on the contrary, they contrasted the fake European democracy to the Tunisian revolutionary experience: 'How is it possible that we are in Europe, the supposed place of human rights and we are left in these conditions? Is this Europe? We want just to remind that we hosted thousands and thousands of people escaping from Libya, and Italians are not able to solve this situation'.[13]

A POLITICS OF DIS-CHARGE

Thus far, I have illustrated the substantial recalcitrance of migrations taking place in the aftermath of the Arab uprisings in fitting into the terms of political recognition and representational politics. Hinging on these considerations, I examine the ambivalence that is at play in the effective mechanisms of the migration. I would call this ambivalent tactic a 'politics of dis-charge', meaning by this a technique of government grounded on two simultaneous moves: on the one hand, the politics of control and humanitarian assistance, and on the other, the politics of indifference, whereby people are left stranded without assistance and at the mercy of fate (discharging). In other words, what I question is the idea that migrants are subjects who are governed more than others, assuming the notion of government as defined by Foucault but less in its more formal meaning—government means to structure the field of actions of others (Foucault, 1982, 1997)—than in its pastoral version of a government of individuals' conduct and of the population at the same time, based on a logic of 'salvation' (Foucault, 2009).

In addition to taking *in reverse* and *by surprise* the discourse of migration government as geopolitical phenomenon to manage, we should look at the *reshuffling of identities* produced by the migration regime: indeed, practices of movements are labelled, fixed and spatialized through 'mobility profiles' and 'legal geographies' (Basaran 2011). Categories that, as I will explain in chapter 4, at times are fully assumed and then reversed or altered in their function by migrants themselves. However, if embracing the same categories through which people are governed by power could be tactically useful in some specific moments, the moment when we should try to depart from those identifications is one of the main political stakes that Foucault suggests to address. In fact, referring to the feminist and homosexual movements of the 1970s, the French philosopher notices that their political force lay not only in their claim to be recognized for their sexual specificity, but in disengaging from the very normative frame of sexuality and trying rather to instantiate new forms of relations beyond juridical contracts.[14] These movements invented and put into practice new forms of sociability and existence beyond the codes and the boundaries of the dispositive of sexuality. The category of 'migrant' is certainly the primary mobility identity to challenge, both by those who write on migration, and by those who are labelled 'migrants'. If the question 'who is a migrant?' undoubtedly poses a difficulty for the normative categories of migration, once we try to answer this question, we realize that it is almost impossible to abstract from the situatedness of specific times and contexts; and it becomes clear that the 'migrant-subject' is the provisional outcome of governmental technologies and spatial captures that craft legal subjectivities and also effectively produce some people as 'migrants'. Therefore, the migrant condition in its heterogeneity of forms and singular experiences is shaped in time and across different spaces of governmentality—the space of the frontier, the space of the camp, the space of the harbour where migrants from Tunisia disembark, the space of the embassy.

The focus on the migration turmoil in the Mediterranean at the time of the Arab uprisings enables us to break and multiply the monolithic notion of migration that is usually taken as an indistinct signifier for designating any kind of difficult mobility. On the contrary, the blurred signifier of migration must be broken apart according to the heterogeneous practices of movement. This doesn't mean to reproduce the exclusionary partition between forced migration and voluntary economic migration; rather, it is a question of recognizing the differences concerning the reasons and the conditions of people's mobility, opposing the epistemic violence of the overwhelming catchword of 'migration' in the face of the materiality of diverse living conditions, desires and power relations. This clarification should be kept in mind also in the debate about autonomous migration: indeed, the risk of making general assumptions about the 'excess' of migration in relation to border controls is to take specific practices of migration as paradigmatic of any migration, thus disconnecting the analysis from the concreteness of the different strugglefields of governmentality. In the case of Tunisian migrations this emerged quite glaringly, and the difference should not necessarily be framed in terms of forced and unforced migrations. Actually, it could be rephrased in multiple ways corresponding to the diverse migrant conditions: the relation with the country they left—people arriving in Italy from Libya were all migrant workers in Libya, while Tunisians were citizens in their country—the desires at stake—finding a safe place to settle or travelling in Europe—and the political context they left—the revolutionary and popular uprisings in Tunisia, the NATO conflict in Libya.

This difficulty confirms the historical fluctuation of the meanings and of the objects of migration government: who is 'made up' (Hacking 2004b) and governed as 'migrant' changes over time. Thus, we could try to sidestep, or better to reformulate the question through two other questions. Who today in this context is 'migrantized' (i.e., has become migrant) in the face of the current processes of impoverishment? And, jointly: who, today, in this specific political context, is labelled and governed as a migrant by migration policies? In other words, the geopolitical location, the mechanisms of precarization and the political inscription of bodies into mobility profiles need to be rethought together. The sorting and labelling processes enacted by migration policies have the power to 'transform' practices of mobility into migrations. Simultaneously, we should consider the migrant also as subject to a social and economic condition related to an array of processes of precarization that bring about a forced and restricted mobility. Indeed, it is by articulating these two lines of interrogation that it becomes possible to disconnect the normative force played by the category of migrant without losing its political value.

Taking together Marx's analysis on the production of a population in excess and Foucault's consideration of the emergence of a norm as a set of habitudes and mechanisms of constraint that marginalize and banish 'dangerous conducts', we can interrogate the space of governmentality open up when these mechanism work together (Marx, 1993; Foucault, 2013a). This becomes quite visible if we take strategies of migration as a standpoint: undocumented migration is in fact a strategy for constantly trying to dodge and subvert the 'abnormality' of their unauthorized conduct, the condition of being an 'excess' of productive/reproductive mechanisms as well as of free mobility, and, jointly, the 'abnormality' of being out of place. That is, the detention of 'irregular' migrants works as a sort of decompression chamber for introducing cheap labour into the market. But at the same time undocumented migrants are usually deprived of any access to the means of reproduction—constituting in some way a population in excess—and they are counted out from any selected politics of mobility. However, the migrant condition also corresponds to the fact of being destituted as subjects of right, as individuals who stepped out from the 'citizen pact' of their country and are fundamentally out of place in relation to where they are expected to stay, recalling Foucault's idea of an infra-legal illegalism (Foucault 2013a): the 'scandal' of unauthorized migration consists less in the transgression of law than in the attempt to subtract from a regime of life and to enact freedom irrespective of geopolitical inequalities. Thus, by assuming migrant struggles as a standpoint it enables us to shed light on the mechanisms of government over life that operate through dis-charge, combining the production of 'irregular' subjectivities and the multiplication of people in excess.

THE MIGRATION-GOVERNMENT NEXUS: WHICH CRITIQUE?

'Migration' has today become a sort of self-evident catchword entangled with multifarious social and political issues forming a sort of dodgy continuum connecting terrorism, insecurity, poverty and trafficking, and a humanitarian-securitarian bond. The underlying political and epistemic linchpin that sustains the translation of some practices of mobility into migration is the paradigm of government. Migration as a social issue is coupled from its very outset with a governmental rationale: migration is a phenomenon to manage. And, given the regulative function of borders, it follows that there wouldn't be migration outside of a governmental perspective. 'Government' stands in the migration fields also for the epistemic blueprint that shapes our way of thinking and speaking of people's movements: some practices of mobility—migration—are postulated as objects of government, since they alter the ordinary political and economic stability. Most critical analyses on migration management ultimately assume migration and government as an unquestioned couple, addressing criticisms of inhumane treatment of migrants or against the deprivation of migrants' political and civil rights, insisting on the necessity to grant access to services and protection. In this regard it is worth recalling Foucault's distinction between critique as a judgement or a denunciation and critique as inseparable from transformation and as its condition of possibility (Foucault, 1994e). Therefore, critique does not consist in saying what is wrong but in refusing and resisting a certain mechanism of power: 'There is not a time for the critique and a time for transformation; there are not those who need to criticize and those who have to transform. . . . The work of a deep transformation cannot be done other than in the free and always agitated air of a permanent critique' (Foucault, 1994g: 457). Actually, critique can be conceived as a critical attitude: in this way, critique is primarily a posture 'which consists in seeing on what types of assumptions, of familiar notions, of established, unexamined ways of thinking the accepted practices are based. . . . Understood in this sense, criticism is utterly indispensable for any transformation' (Foucault 1994b, 456–57). Once posited in these terms, what remains is how to link the critique of the technologies of profiling with a possible struggle against them. Concerning the government of migrations, this could be translated in the attempt to speak of practices of movement trying to break the chain of partitions (economic migrants/refugees/clandestine migrants/denied refugees, etc.) and the chain of equivalents (illegal migration = phenomenon to govern = dangerous mob) (Laclau, 2007) that from time to time is reproduced. Such a critique could engender some crisis in the reiterative mechanism of the world making of migrations (Walters 2013)—that is, in the discursive and non-discursive production of the 'world' of migration as an object of government. Such a critique shifts from the question 'who is a migrant?' to the twofold question, always historically and spatially located: who is designated as a migrant, here and now, by migration policies? And who has been migrantized by processes of precarization? That is, we need to speak of practices of movement without translating some of them immediately into migrations (De Genova 2013b). But practices of migration sometimes succeed in making border interruptions, without necessarily transforming or challenging those boundaries. To put it concisely, migrants are not interested in exploding partitioning categories or in subverting the discursive regime, but rather in moving or persisting in certain spaces. By enacting practices of mobility, migrants disrupt the spatialization of people's movements that migration policies put into place: but they misdirect it not because they contest its mechanisms or because they enact extreme forms of resistance, but for the simple fact of moving or staying. However, I do not conceive here interruption as an event which suddenly breaks the political space, creating a new order. In fact, the concept of interruption is posited as a central tenet in many philosophical analyses on subjects' agency and on the ways in which borders and norms are deeply challenged by unexpected acts of discourses enacted by those 'who are not covered by those norms or have not entitlement to occupy the place of the "who' (Butler 1997, 367; see also Isin 2008, 2012; Laclau 2005; Ranciere 2004a, 2004b). But the association between interruption and event depends on how we conceive the event itself, and in this regard Foucault's perspective differs from messianic formulations as well as from the idea of a punctual emergence of subjects on the political scene of address. Indeed, as Foucault explains 'it is not a decision or a fight but a rapport of forces that overturns, a confiscated power, a vocabulary that is appropriated and turned over against its users. . . . The forces at play in history obey neither a destiny nor a mechanics but the hazard of the struggles' (Foucault 1984a, 1016). In this wake, interruptions relate to unexpected movements which upset and exceed also the 'legal production of illegality' (De Genova, 2004), disrupting some mechanisms of capture from the position of subjects governed by that regime.

Independently from the degree of friability of the migration regime, those who are labelled 'migrants' come up against mechanisms of capture and selection. This is a discursive order which traces a *moral geogra-*

phy by sorting people into exclusionary channels of mobility and corridors of layered protection. In a nutshell, migration policies work through the fixation and dislocation of bodies in space. Starting from these assumptions, it is fundamental to disengage from any reformist or humanitarian approach regarding spaces of detention—criticizing the inhumane conditions in which migrants are treated. In fact, the administrative arbitrariness of migrant detention centres should not be seen as a deviation from an ideal global migration regime but as its effective 'economic operator' (Foucault 1980a, 136). Finally, if the mechanism worked according to the policy textbook, there would not be 'illegal' migrations at all. The question raised by Foucault regarding the prison and the penal system—'is it possible to have a society without illegalisms?'—provides a productive insight to start a political and critical engagement about the function of migrant detention centres. Therefore, from such a standpoint the very existence of prisons for people not authorized to move should be framed through an investigation of the economy of illegalisms, focusing on the distance that, ultimately, the detention of non-citizens introduces between the 'illegal' position of undocumented migrants and the criminals as breakers of the social contract. Indeed, undocumented migrants are definitively outside the social-contract while, on the contrary no citizen cannot be 'illegalized' for their mere presence in national space.

Therefore, what comes after (and alongside) the critique? And jointly, what should a critique trigger? Rephrasing Foucault's reflections on the struggles over prisons, it could be stated that the goal is neither to make the prisons visible nor to make detainees aware of the injustice of the mechanism of detention, but rather to make the detention system intolerable and to situate it in a broader economy of power which needs to be challenged.[15] These considerations lead us to ask from what position and what ways can one resist techniques of bordering. As Foucault underlines, resistances to a certain regime of truth cannot come but from the inside of that regime, namely, by those subjects who are produced by that system of veridiction (Foucault 2009, 2010). However, one could object that many of the techniques of surveillance adopted to control migrant's movements are also employed to monitor and manage, in a different way, the mobility of the 'non-migrants' (Bigo 2005; Pickering and Weber 2006). Also for this reason it is important to maintain specific claims and rights for migrants, trying rather to restate that the inequalities and violations experienced by migrants are in fact a lens for pointing to the mechanisms of precarization that involve both citizens and non-citizens. Nonetheless, a distinctive factor can be identified: migratory policies have a concrete impact on the existence that people could effectively live once they are labelled 'migrants' by states and governmental agencies. From this perspective, it is necessary to reiterate the question: 'who has become/is becoming a migrant here and now?', taking 'migrant' in the double meaning of subject of migratory policies and subject of the real process of migrantization. Challenging, interrupting, criticizing, infringing, escaping and subtracting: all these verbs signal strategies of resistance and different modalities of counteracting that we find at play in the forms of struggle and existence of and about migrants. To what extent can a critique of the government of migration, as radical and troublesome as it might be, be effective in producing interruptions in the functioning of migration governmentality? Indeed, an interruption cannot be generated other than from within the regime of truth, and consequently by the side of those who, from time to time, are labelled as 'migrants' or are migrantized (Plascencia 2009). To put it differently, for a temporary short-circuit to be produced, someone has to be in the condition of striving for escape, subtracting from that regime because of the unbearable impacts on his/her life. Instead, critique does not imply that one resists, escapes or intends to do so: rather, the critical gesture involves keeping a distance.

This consideration suggests that one is a migrant when, in order to move or stay in a place, one needs to resist or dodge the politics of control. Take for instance the airport, a site of intense control where the fact of everyone being governed by a global regime of mobility emerges clearly; however, at the same time, it makes visible the unequal distribution of the *grip* of the politics of mobility as well as the different degrees of tolerance towards controls: the ordinary gestures of showing the passport to national authorities or passing through the body-scanner are accepted by bona fide travellers. In other words, people who have all the papers to move legally shift from a critique of the intrusive character of controls and their desirability as dispositives that make freedom possible (Bigo 2011; Foucault 2009; McGrath 2004).[16] The ambivalences of the desirability of controls should also be considered: controls appear not only useful but also desirable when it becomes a concern of public security, when controls are applied for sifting and selecting travellers in order to prevent acts of terrorism. If we refuse to take on migration as a domain per se, governed by exceptional self-standing laws, and we situate it rather in the global labour regime, it becomes perceivable that an ease of circulation in space is not necessarily an index of a (high) degree of freedom and autonomy from the government of mobility (De Genova 2013a; Mezzadra 2006). The relative ease of mobility of high-skilled migrants all over the world is enhanced by economic actors; and thus, it is not so evident that less obstructive conditions of movement and the 'softness' of controls stand for much freedom from mechanisms of exploitation and regulation. As Didier Bigo remarks, 'under liberal governmentality, mobility is translated into a discourse of freedom of circulation, which reframes freedom as moving without being stopped . . . freedom has often been reduced to freedom of movement' (Bigo 2011, 31). Consequently, the paradigm of circulation could be misleading as a description of the economy of interrupting movements—and thus, of producing immobility—that characterizes the government of migration. Rather, 'who has access to circulation?' and 'under what conditions?' are the questions which should drive our critique of the migration regime (Aradau and Blanke, 2010). In fact, circulation flattens migration on a smooth space of managed mobility, while it actually consists in disrupted movements: migration is ultimately a hindered form of mobility, made up of thwarted projects of life and movement (as peoples' journeys and projects being literally governed by migration policies).

POLITICS OF NON-TRUTH AND 'GOOD' STORIES: THE GOVERNMENT OF WOULD-BE REFUGEES AT THE LIMIT OF THE WORLD'S HISTORY

If we assume that migration governmentality is at the same time a laboratory and a litmus paper of wider economies of power over life, it should be investigated whether the governing of conducts and the production of subjectivities always work mainly through a regime of truth which requires subjects to produce a discourse of truth (about themselves). Or, rather, if in some spaces and strugglefields of governmentality there are also other regimes at play that do not postulate a subject of truth. In the inaugural conference of the lectures held in Louvain in 1981, Foucault stressed the proliferation of truth telling as well as the multiplication of the regimes of veridiction in our contemporary societies. In that case, the pronoun 'our' addressed the European space and its corresponding form of subjectivity. However, despite the plurality of the regimes of truth mentioned by Foucault, his genealogical account concerns the emergence of the regime of truth underpinning the production of the modern Western subject. But in order to grasp the emergence of a spatially and historically situated subjectivity it is necessary to intersect both the regime of truth and the process of subjectivation against which that subjectivity has been shaped.

A genealogy of the modern European citizen might be traced both from the time of the constitution of national citizenship and in moments of economic or social crisis, when the unity of the working classes is replaced by an exclusionary politics by which the immigrant is seen as an undeserving recipient of social benefits. It has been widely demonstrated that the very figure of the immigrant, as distinct from the foreigner, came to the fore along with the consolidation of the citizen's identity (Noiriel 1996, Wahnich 1997). Nevertheless, although a genealogy of the contemporary European subject requires complicating it with processes that craft subjectivity from the 'outside' of that space, it is not merely by refraction and against border identities—like the immigrant and the colonized—that the emergence of the citizen-subject should be grasped. Rather, what is important to scrutinize is how techniques of identification or of categorization devised for managing migrations are reorganized beyond their original scopes, becoming central tools for shaping and governing citizens' mobility. More precisely, differentiated degrees of citizenship can be instantiated just through the production of inner 'lacking' and 'border' subjectivities. Looking closely at the governing of 'untruth conducts' also makes it possible to grasp transformations taking place in the political technology which governs non-migrants. That is to say, by situating ourselves at the edges of the regime of truth that is supposed to shape the space of citizenship we enable the foregrounding of reconfigurations of power that are at stake also within that space.

As Ann Laura Stoler remarks, questioning the proliferation of discourses on sexuality in our societies stressed by Foucault, these discourses 'were refracted by men and women whose affirmation of a bourgeois self was contingent on imperial products, perceptions and racialized others' (Stoler 2002, 144). And the coexistence in present societies of subjects whose discourse of truth responds to very different injunctions prompts questioning the pronoun 'our'—referred to in 'our space'—bringing out the overlapping of heterogeneous but co-interrelated regimes of truth and subjectivation in which we are simultaneously entangled. To put it differently, we could rethink how mechanisms of truth's production are co-determined in such a way that a history of our present cannot carve out autonomous and separate regimes of subjectivation. The plurality of regimes of veridiction goes with the differentiation of spaces, suggesting the inevitable plurality of histories that a history of the present involves.

This digression on the importance of complicating the genealogies of Western subjectivity leads to technologies for governing conducts that although located in our present are assembled through dissonant regimes of veridiction. The governmentality of refugees and the blurred lines between politics of protection, government of the undesirables and detention of undocumented migrants, constitute topical sites and mechanisms to tease out the coexistence of different regimes of truth and to see where the very issue of subjectivation through individualization—that is, through a discourse of truth of the subject upon itself—is at least partially unsettled (Vaughan 1991).

In order to go more deeply into this issue, I turn to a place at the edges of the Mediterranean space,[17] the Choucha refugee camp at the Ras-Jadir frontier post on the Tunisia-Libya border. Limits here invoke the constant production and realignment of borders insofar as a process of denomination takes place—in this case the Mediterranean space of free mobility—in the name of shared historical/cultural legacies. But beyond the geographical location, I focus on the Choucha refugee camp to analyse closely which mechanisms and obligations of truth telling are at stake in governing would-be refugees. Choucha is not a special place but, on the contrary, it allows us to highlight the politics of (non)truth at play in governing refugees' conducts: thus, I take it as a trans-local space for looking at the functioning of certain regimes of veridiction. From this standpoint, Foucault's genealogy of the technique of confession and its contemporary transformations—psychiatry and the juridical system—work here as a set of analytical coordinates through which to read both the regime of truth and the production of subjectivity at play in the domain of governing refugees. However, the scrutiny into the production of truth in the government of refugees raises issues far beyond the specific field of migrations, disturbing the supposed self-standing genealogy of the Western contemporary subject.

Therefore, the politics of (non)truth at stake in governing would-be refugees operates here both as a litmus paper of a broader transformation of power and as a laboratory of governmentality bringing into existence new or transformed political technologies.

GOOD STORIES AND THE CONFESSION WITHOUT TRUTH

December 2011: 'We are aware that we are not asked to tell the truth, to tell our truth story, rather we need to tell a good story'. People who fled from Libya in 2011, most of them third-country nationals working in Libya and forced to leave after the outbreak of the war, know this. It is not the truth—that is, the effective events that occurred in the lives of those would-be refugees—which really matters for UNCHR's commissioners, who decide whether to give them a space on the earth where they can legitimately stay. In a sense, there is an 'excess of the real in their stories that cannot be symbolized' or grasped by juridical categories (Beneduce 2008, 507), because it doesn't refer to any deferrable or 'in place of' meaning to unfold, but rather it addresses lived experiences. But what are the 'good stories' that the refugees mention as stories to be told? (Good 2004). Having in mind Foucault's description of the obligation for the subjects to tell the truth upon themselves, in the case of governing would-be refugees, the speech of the migrant is postulated as untruthful in principle, or better, as suspected of being lies: 'this systematic suspicion regarding the asylum seekers transforms the inquiry on truth telling into a process of lie detecting' (Fassin 2013, 54).[18]

The reason why would-be refugees refer to a 'good story' to tell as the only possibility to get international protection is that the discourse required by the territorial commission which processes asylum claims essentially must comply with a set of normative categories and conditions (Zetter, 2007). The discourse of the would-be refugee is supposed to comply with a truth that in some way is already there—the truth actualized in profiles of mobility—according to which people are spatially re-located; then, the primary partition of economic migrants/refugees is multiplied into a differentiation of conducts corresponding to a downgrading continuum going from protection to unprotection (Bohmer and Shuman

2008; Squire 2009): vulnerable person, 'ordinary' refugee, suspect-fake refugee, denied refugee. Recalling Fanon's considerations on the treatment of the colonized, what is important to notice is that the various conducts of would-be refugees emerge only through the clash with governmental truth and power (Fanon 1994b; Foucault 1994a): 'the North African does not come with a substratum common to his race, but on a foundation built by the European' (Fanon 1994b 7). In fact, UNHCR's procedure for assessing asylum seekers is grounded both in the 1951 Geneva Convention and a scheme, much more variable over time, listing safe and unsafe countries. However, this geopolitical map of unsafety is not the only condition upon which the decision is made: actually, the articulation of personal stories and nationality is what in principle forms the ground for examining the reasonableness of asylum applications. And the yardstick for assessing the reliability of asylum seekers' stories depends fundamentally on two criteria: the contradictory dimension and the inconsistency of the would-be refugee's discourse, with the latter ultimately prevailing over the former. To sum up, there is no discourse of truth that migrants and would-be refugees are considered able or willing to utter. Their speech is judged on the basis of what I call an asymptotic adherence to the moral cartography of the regime of asylum (Malkki 1992): the national origins and the degree of vulnerability of the asylum seekers define the two main coordinates through which the mobility profile of the would-be refugee is shaped. Ultimately, the examination process hinges on a sort of *defensive proof*. Would-be refugees need to demonstrate acceptability against the *moral geography*[19] traced by UNHCR and designed to label them as non-eligible for protection, providing evidence for questioning the geopolitical narrative that encapsulates their stories into a landscape of spaces and subjects at risk. Indeed, asylum seekers have to demonstrate that their story is an exception to the geography of 'safe/unsafe country' established by UNHCR.

UNTRUTH CONDUCTS AND THE MORAL GEOGRAPHY OF ASYLUM

Nevertheless, despite the mark of non-credibility that sustains the discourse of impossible truth demanded of the *non-refugee until proven otherwise*, what is at stake is not a silence about their lives. On the contrary, the government of migration is characterized by a high rate of discursivity through which would-be refugees are governed and encoded into an intelligible and standardized schema of conducts. Thus, despite the fundamental discrediting which underpins the speech of would be refugees, it remains that an injunction to speak, to tell one's own story, pervades the technology of government of displaced people: the asylum seeker is demanded to tell not only the story of his/her journey but of his/her entire life. However, what predominates is not the truth of the asylum seeker's speech on his/her life but, rather, the truth of his/her geopolitical location and of his/her singular story as part of that moral geography: the truth is produced irrespective of the discourse of the would-be refugee. But this doesn't mean that the story told by the suspect subject is irrelevant: on the contrary, the storytelling works as a sort of normalizing technology demanding that the migrant comply with the geopolitical narrative provided by international agencies, humanitarian actors and states. At the same time, the subject could give up on the impossibility of telling the truth, narrating instead the 'good' story to the Commission and arguing that his/her vulnerability definitively makes an exception to the rule. Therefore, it could be named a *confession without truth*; a quite odd practice of confession that does not postulate any hidden thought to unfold but, rather, posits an already-there reality envisaged by international criteria and treaties, which the subject is demanded to embrace. Unlike the correspondence between discourse and subjectivity that the psychiatric confession requires of the individual (Foucault 2012a, 2012b), in the government of would-be refugees it doesn't matter that subjects authenticate themselves and their subjectivity by attaching themselves to the discourse they formulate upon themselves. Instead, what is at the core of this regime of veridiction is exactly the dissociation between the speech of untruth of migrants and their juridical status fixed by normative international criteria: the notion of truth is partially superseded by the idea of a statement of facts, namely, an objectivity that stems from an indisputable evidence graspable through standardized knowledges and practices of expertise. By scrutinizing the effects of subjectivation generated by this confession without truth, what emerges is a growing array of 'profiles of mobility' that singular stories must fill in and that get troubled when juridical categories become unsuitable in keeping up with the heterogeneity of migrants' practices. UNHCR acknowledges that in many cases both migrants' conditions and practices of migration do not fall into existing mobility profiles and migration categories. Profiles are created both for individual conducts and for countries, proving that humanitarian government lies at the junction between a moral geography of conducts and a governmentality of scattered populations where nation-states still play an important role.

To sum up, in the governmentality of would-be refugees, migrants' speech is disqualified from the very beginning, and no discourse of truth is supposed to be formulated by asylum seekers. In contrast to that, it should be stressed that the main function of confession, also in its secular forms, is to attach the subject to his/her own truth; and this truth is not coming from the outside but it coincides with the very discourse that the subject is forced to produce upon him- or herself. Instead, if we shift to the examination process of would-be refugees, 'assessment of credibility' is the buzzword extensively used by humanitarian agencies for postulating the discourse of the migrant as an in-principle contradictory one; at the same time, would-be refugees are entitled to the right to defend themselves against the evidence of facts and statements formulated in reference to their life/story and to their geopolitical location.

WHEN CATEGORIES DO NOT WORK ANYMORE: THE SUBTRACTION FROM THE DIAGNOSTIC GRASP

Ultimately, the aim of the examination consists of making the subject contradict her- or himself, playing with the incongruities that the examiner finds in the migrant's speech. In this regard it is relevant to recall Fanon's descriptions on the difficulty for psychiatrists to make a diagnosis of the illness of the colonized due to his resistance to accounting for himself: 'The refusal of the Muslim to authenticate through the confession of his gesture the social contract that is offered to him, means that his effective subjugation cannot be confused at all with the acceptance of that power' (Fanon 2011, 126). On the one hand, the colonized refuses to authenticate his/her act, disengaging from the subjectivity to which the diagnostic knowledge tries to bind the colonized; on the other hand, this refusal is staged through an 'orchestration of a lie' (Fanon 2011). Therefore, in some cases diagnostic categories fail to tell the truth about the subject, since this latter at some point resists the possibility that a diagnosis could be made; and this resilience is displayed also at the level of body: 'the doctor would have to conclude that medical thinking was at fault . . . and he finds the patient at fault—an unruly, undisciplined patient, who doesn't know the rules of the game' (Fanon 1994a, 8).

Perhaps, following Fanon's considerations, we could reverse the gaze on the governing of refugees from the standpoint of the governed subjects: in the place of the moral cartography enacted by the governing of untruth conducts, the persistent elusiveness of would-be refugees to make their biographies readable by the regime of veridiction, traces out another map. Indeed, there are biographies that cannot be fixed into profiles or narrated into stories. Or, stories that prove to be an exception to the unquestionable truth as evidence upon which the government of would-be refugees is grounded. Putting in conversation Fanon's analysis with Foucault's considerations on the politics of truth, it could be suggested that when the would-be refugee refuses to sign the 'social contract' offered by the moral geography of governmentality—making impossible-to-translate practices of mobility into profiles—he/she cracks the consistency of that regime of veridiction which seeks to make migrants' journeys intelligible. Nevertheless, it doesn't mean that the subject thoroughly subtracts from that regime; in fact migrants' deportability (De Genova 2010b) and their condition of being stranded, living in a frozen-time dimension, tell us how power carves out their bodies and lives. Rather, despite the meshes of 'diagnostic truth', would-be refugees (sometimes) do not authenticate that regime of veridiction, strategically playing with their untruth conduct and producing a strategic inversion of the confession without truth that is requested of them. Indeed, despite the fact that asylum seekers are demanded to go into detail on the narration of their life, the injunction to tell a good story is not grounded on individualization; and, secondly, I suggest that it seems to not involve a process of subjectivation in the sense illustrated by Foucault. Instead, by forcing biographies into a pre-existent moral cartography of migration categories, they give rise to a new, updated compound of unsettled existences.

According to Foucault, secularized techniques of confession basically function as 'therapies of truth' that the subject is compelled to engage in; in this way it could be stated that the couplet 'coercion-therapy' is at the very core of the injunction to speak and to tell the truth upon oneself. The peculiarity of the modern therapies of truth consists in postulating the dependence of our salvation—secular safety—on the obligation to know who we are: subjects must hold a thorough knowledge of themselves, and such a truth has to be regularly verbalized. This entails that the disciplinary function of the techniques of confession lies in a (true) production of knowledge that takes place through the discursive engagement of the subject. But is this the case also for the good stories that would-be refugees have to craft? At close scrutiny it seems that any therapeutic function is excluded from the discourse of the refugees. In fact, the obligation to make one's own life intelligible and readable doesn't come from the need for self-care or to save the suspect subject, and no 'therapy of citizenship' seems to be possible: indeed, also the politics of resettlement which relocates people in third countries, assigning them a space to stay, is not conceived on the basis of a logic of therapeutic 'salvation', but rather as a biopolitical relocation of bodies and existences in space. Moreover, the asserted impossibility of care depends also on the colonial legacies that still pervade the body and the geographical origins of the would-be refugee: the colonial subject was depicted not only as a suspect subject but also as an incurable body which escapes from all diagnostic categories. The unreality that, as Fanon illustrates, is judged in the colonies as the hallmark of the pain of the colonized underlies also the truth-telling of would-be refugees (Fanon, 2011).[20] In other words, in the absence of any possible therapy of truth, two distinct but overlapping levels are at play: on the one hand, the individual level of conduct—where would-be refugees are assumed to be vulnerable subjects to take care of, but whose latent vulnerability ultimately remains incurable—and on the other, the impossibility of therapy is translated on a global scale, concerning the governmentality of migrants and displaced persons.

Hence, if as Foucault remarks there is a direct implication between the regime of veridiction and the production of subjectivity, what are the effects of a regime of truth that is dislocated from the subject? Paradoxically, the very partition between an inside and an outside of the regime of truth becomes unsettled by would-be refugees' strategies: the production of 'good' stories—that is, the choice to craft and alter one's own story and biography in order to meet the criteria for getting refugee status involves a truth-telling to which the subject ties itself. Or, to the contrary, by resisting any 'diagnosis of truth', would-be refugees undermine the functioning and the tenability of migration categories and mobility profiles. This inroad into the government of would-be refugees has highlighted that would-be refugees are not the mere reversal of the political subject or of the good citizen; rather, they emerge from the primary partition between migrants and refugees and from the multiplication of different degrees of (un)protection, configuring an unstable moral cartography of legitimate or undesired presences. Conceiving the history of the present as an investigation for 'understanding how subjects are effectively tied into and through the forms of veridiction in which they are engaged' (Foucault 2012a, 3), this cannot be made 'by writing in the comfort zone' (Stoler 2002) or by positing a form of subjectivity as a yardstick by which to measure all the others. A critical focus on the regimes of truth at play in the government of would-be refugees makes it possible to bring out the differential mechanisms of 'governing through truth': a gaze on the 'working differently' of the mechanisms of power and truth in the government of would-be refugees could be used to investigate, starting from a marginal and specific location, the 'production of truth' required of the citizen-subject.

NOTES

1. *Storie Migranti*, http://www.storiemigranti.org/spip.php?article1049.
2. A similar move is undertaken by those scholars who mobilize a 'regime analysis' of migrations, looking at the government of migration as a space of negotiating and conflicting practices (see among others, Hess, Karakayali and Tsianos, 2009 and Karakayali and Tsianos, 2010). This work is in part situated in that perspective, assuming migration controls as the effect of conflicts and practices for taming practices of mobility; but at the same time it engages more closely with the Foucaultian notion of governmentality, through which subjectivities are not assumed as what power tries to capture or govern (postulating their substantial autonomy and their being-already-there) but rather as the outcome of specific and strategic games between freedoms and power relations. Thus, more than taking migrants' subjectivities as a starting point, this analysis explores 'the ambiguous position of subjectivity', meaning by that the complex articulation between how it is produced within the strugglefield of power relations, and how it is productive (Read 2003).
3. This is an issue that recurs many times in Foucault's work for describing both uprisings and struggles—as for instance in the case of the writings on the Iranian revolution—and his own work. In an interview of 1978, Foucault defines his work as an attempt to 'displace the forms of sensibility and the thresholds of tolerance'. From this standpoint, also the meaning of critique comes to be radically redefined: 'critique is a challenge in relation to what's there' (Foucault 2001f, 851).

4. The critical reference to rationality is a recurring motif in Foucault, as here he succinctly argues 'the government of men by men involves a certain type of rationality. It doesn't involve instrumental violence . . . so the question is: how are such relations of power rationalized?' concluding that 'political rationality has grown and imposed itself all throughout the history of Western societies . . . its inevitable effects are both individualization and totalization. Liberation can come only from attacking not just one of these two effects but political rationality's very root' (Foucault 1994f, 324–25).

5. In this regard, it is important to notice that, as Karakayali and Rigo contend, despite the intention by European politicians to establish a common regime of immigration 'what has been strengthened was the common administrative body of combating migration' (Karakayali and Rigo 2010, 31).

6. In this regard, Foucault nicely captures the potentiality of gay movements to invent relations without any form, yet; and to multiply the possible forms of relations and modes of existence that spill over the existing (juridical) ones. In particular, Foucault underlines that the main reason why homosexual relations are disqualified and marginalized in our society depends on the unpredictable social relations that they could create. My hypothesis is that in these analyses Foucault shifts a little from the issue of the conduct—how to conduct myself in a different way—focusing rather on processes of subjectivation and potential relations to create that are immediately also collective; indeed they do not start from a relation to the self and then open up a common dimension, but rather shape from the outset multiple relations and put into place collective ways of life, that sometimes could lead to what Foucault calls a subculture. Secondly, the stress on virtual new relational schema and practices as well as on the invention of unexpected relations among individuals draws the attention more on mechanisms of subjectivation-desubjectivation than on relation of mastery and autonomy over oneself.

7. Addressing neoliberalism, Aiwha Ong contends that 'new forms of governing and being governed and new notions of what it means to be human are at the edge of emergence' (Ong 2006, 4).

8. The framing of migration government in terms of 'management' is quite recent, as Geiger and Pecoud (2010) point out: 'the notion of migration management was first elaborated in 1993 by Bimal Ghosh following requests from the UN Commission on Global Governance and the government of Sweden. In 1997 the United Nations Population Fund, together with the Dutch, Swedish and Swiss governments, financed the so-called NIromp project (New International Regime for Orderly Movements of People). . . . The idea was that, in the post Cold-War era, migration had the potential to generate real crises and that a global and holistic regime of rules and norms was needed to successfully address the phenomenon'.

9. The notion of 'strugglefield' designates the strategic configuration of power relations and resistances: it frames the very relationship between power and resistances not as a dynamic of action/reaction but as a confrontation between forces, and in this way power is nothing but the present and unstable 'winning strategy'. This implicates that (a) resistances cannot be but internal to power relations and that (b) powers and resistances relate to each other according to a sort of permanent limit [Foucault, 1978b]. Secondly, the notion of strugglefield conceives of governmentality as a conflicting space in which, as Foucault contends, the interrelation between government of the self and government of others is precisely what makes it always possible to find spaces of resistances and points of fragility to invert, transform or break the existing configuration of power. In fact, the notion of strugglefield foregrounds the idea that subjectivities are not eclipsed in the concept of governmentality but, rather, are really at its core in the double meaning of 'subject' (being subject to and being subject of).

10. Derrida's critical analysis about the implications of structuralist thought that risks leaving spatial metaphors and geometrical figures superseding the qualitative

and intensive components suggests not reducing spatial perspective on migration—that this book overtly undertakes—to a cartography of borders and movements.

11. Tunisian migrants arrived on the island of Lampedusa in February 2011.

12. A special six-month temporary permit, for 'humanitarian reasons', was given by the Italian government on April 5. The French government refused to recognize it as a valid document for allowing Tunisian migrants to cross the French border, giving rise to a fundamental quarrel not only between France and Italy but also at the European level, raising proposals for revising the Schengen area as a space of free mobility.

13. Interview with Tunisian migrants in Paris, November 2011.

14. On the concept of relational right see, *The social triumph of the sexual will. A conversation with Michel Foucault*: with the term 'relational right' Foucault talks about the possibility to structure within a certain institutional field some social relationships that do not necessarily refer to the emergence of a recognized minority group. And at the same time, the term designates social relations that cannot be encoded into a juridical frame: indeed, according to Foucault, contemporary societies are characterized by a fundamental impoverishment of the social fabric, narrowing all human relations to juridical ones (Foucault, 1994f).

15. As Foucault put it, talking about the activity of the GIP, 'our inquiry does not aim at gathering knowledges but to increase our intolerance transforming it into an active intolerance' (Foucault 2001c). In short, the very strategic positioning of the GIP consisted not in raising awareness about how power works in the prisons or in denouncing its arbitrariness, but rather in spilling over the prison itself, overstepping the boundaries of that site stressing how that political technology permeates many other spaces. This practical-political positioning is related to what I call 'a movement-towards-the-outside' of the prison itself.

16. As Didier Bigo stresses, 'The advantage of smart surveillance is that . . . for these normalized individuals it would seem to be less of a problem. They appear to be free so long as they do not see those who are controlling their movements, so long as they are not stopped during their journey. . . . This attitude in accepting surveillance is related to this sense that comfort is as important as freedom . . . reassured that they are like a community of travellers where all bad apples have been prevented' (Bigo 2011, 41–46).

17. The very geographical referent of Mediterranean should not be taken for granted, and in particular the boundaries which delimit that space work as exclusive-exclusionary frontiers that separate those outside the Mediterranean area from the cultural and political proximity that Mediterranean countries are supposed to have. In other words, it's in the name of proximity that European countries push for Politics of Neighbourhood with the southern shore and for politics of externalization. At the same time, it implicates that 'non-neighbourhoods' are excluded from such a discursive and political frame. Obviously, the tracing of the border that posits where the Mediterranean space ends, has changed over time, and the Arab Spring was seen as a reason—or as a hope—of getting the Maghreb Countries closer to the political economy of the EU.

18. As Didier Fassin points out, the restrictiveness of asylum that has been in place since the late 1990s has been accompanied 'by a profound loss of credibility of asylum seekers within the institutions in charge of assessing their applications'; and 'the generosity that prevailed in those years [the 1960s and 1970s] was largely a consequence of economic needs for the reconstruction of Europe and the growth of North America' (Fassin 2013).

19. Through this expression I mean the 'geography of the humanitarian' that UNHCR put into place, making a secret list of 'safe countries', whose citizens are considered not in need of protection. More broadly, the expression 'moral geography' refers to the set of criteria according to which would-be refugees are 'allocated' in spaces or resettled in third countries.

20. 'The North African's pain, for which we can find no lesional basis, is judged to have no consistency, no reality. Now, the North African is a man-who-doesn't-like-work. So that whatever he does will be interpreted a priori on the basis of this' (Fanon 1994a, 6).

TWO

Troubling Mobilities

Migrants' Discordant Practices of Freedom and the Power's Hold over Time and Life

> So, through the advanced monitoring systems like the radars you have, you can ultimately see every object at sea and in fact no boat could slip away from your sight.
>
> To see everything . . . it depends what you mean by 'seeing' and what you do with that. Actually, radars reflect fishing boats and bodies into green dots; then, no technological tool could tell you what these dots are the image of[1] .

The quite surprising answer given by the captain of the Italian military corps, Guardia di Finanza, to my questions on the visibility spectrum of the monitoring technologies in the Mediterranean well illustrates the holes and the grey areas upon which the politics of mobility of the migration regime is predicated. Indeed, the rationale and discourse of an exhaustive governability of migration crumble in the face of the fundamental elusiveness of movements which need to dodge the bridling mechanisms of control. And the dramatic paradox of those practices of freedom which evade the governmental grasp over migrants' lives is that such elusiveness means at the same time the impossibility of being rescued. This could be called the migration circle of the government through non-government: that is, the zones of invisibility—something escapes from monitoring eyes and the mechanisms of discharge, leaving migrants dying at sea—through which people's unauthorized movements are captured and channelled. This snapshot of the grey areas of the migration regime illustrates the tipping point at which a critical analysis of governmentality should turn into a substantial reversal of the gaze on migration:

a gesture that in some ways consists in going from the attempt to *map otherwise*—that is, to criticize and challenge the discourse of migration government—to the production of *another map*. In fact, the focus on the revolutionized Mediterranean space after the outbreak of the Arab uprisings brings to the fore the need to envisage 'another' map for grasping the migration strugglefield beyond the grid of government. From a counter-mapping standpoint, 'another map' refers to migrants that sometimes trace and perform unexpected geographies that cannot be encoded into the cartography of government. And, as I will explain later in the chapter, in order to trace their own map, migration controls and policies need to spy on and hijack migrants' knowledge of containing and capturing the troubling mobility.

This chapter explores the reactive character of the politics of mobility as a set of complex responses to the turbulences of migration. Then, it comes to grips with the tipping point of the migration-government nexus, trying to find the limit lines where government as an analytical tool fails to account for the upheavals and constituent movements of migrations. It is precisely by hinging on these limits that government as an analytical grid for reading power relations should be rethought together with the issue of the autonomy of migration.

Critical analyses that want to engage in a reversal of the gaze's reversal on migration, challenging the assumption that migration is a phenomenon to manage, have to be cautious about fully adopting government as an analytical tool. Indeed, the risk is to overlap two distinct meanings (and uses) of government in the field of migration. On the one hand, 'government' indicates the rationale and the political technology that is at play in the mechanisms of control and capture against migrants' movements. And at the same time, the notion of government assumed in Foucault's terms of an action upon the possible actions of the others also provides a perspective slant on the migration regime that makes migration appear as a strugglefield (Foucault 1982): it puts into motion the stable grid of migration studies that is fundamentally grounded on a methodological nationalism, displacing the gaze from the inside/outside of the State to a field of contested and resisted mechanisms of capture that try to bridle and capitalize on those movements labelled as 'irregular'. On the other hand, government tends to be used as a descriptive framework that in some ways comes to overlap and reiterate the narrative upon which those mechanisms of control over mobility and lives are predicated. Government is at the same time the object of a political epistemology that aims to retrace its emergence as an unquestioned language to speak of migration and, at the same time, challenge its legitimacy.

Keeping this in mind, I focus on Tunisian migration taking place in the aftermath of the Tunisian revolution. What distinguishes these practices of migration from others is clearly the revolutionary context and, most importantly, the fact that they have driven the political uprisings out of their national location—Tunisia. Ultimately, the argument that struggles for democracy in Tunisia and the movements of people who left Tunisia cannot be taken separately stems from the idea that it is not possible to speak of effective practices of freedom and real democracy without freedom of movement. On the one hand, this is fundamental for highlighting the uneven democracy envisaged by European states as a model for Neighbourhood countries—namely, an odd democracy which does not contemplate the possibility of freely moving and leaving one's own country. But simultaneously the context of the revolution has further complicated the picture. Indeed, the radical transformations which took place in the time span of a few months—involving the redefinition of the relation between public space, politics and religion—contributed to shaping a quite pervasive discourse, mostly on the left wing of Tunisian society, about the importance of not leaving the country in order to continue the political struggle there. Consequently, the entanglement between revolutionary uprisings and strategies of migration and the possibility to see the latter as an actualization of the former is to be found at the level of practices of freedom but at the same time it remains a very ambivalent and slippery concern.

Tunisian migrants through their flights did not only escape the border regime; they also took and enacted their freedom as freedom of movement, irrespective of the conditions established by the selected politics of mobility—that for instance fixes the terms to get a student visa so that they cannot be obtained by most young Tunisians—and arrived in Italy as a sort of troubling effect of the revolution. Moreover, they played out an uneven (in)visibility, 'at intermittence': after arriving on the island of Lampedusa all together in the span of few months, and thus becoming the 'border spectacle' (De Genova, 2013a) par excellence, those who succeeded in escaping detention centres vanished in many Italian cities, trying to remain clandestine.

THE PRODUCTIVITY OF THE MIGRATION REGIME: PRECARIZATION, UNEVEN CONDITIONAL SPACES AND INTERRUPTIONS

By taking the migration regime in reverse and by surprise, as I proposed in the first chapter, critical analyses should disrupt the solidity of any border regime. We must find ways to account for the frantic and haphazard manner in which migration policies and knowledges seek to come to terms with spatial and political mess. Pushing this perspective forward, taking the governmental frame *in reverse* and *by surprise* allows us to see the *politics of pillage* which sustains migration governmentality: knowledge production on migrations is based on studying, capturing and hijacking knowledge of migrations in order to invent new mechanisms of capture and for anticipating migrants' border crossing (Karakayali and Tsianos 2010). The map of migration governmentality is always a *responsive cartography*: a map that pillages subjugated discursive and non-discursive knowledges corresponding to migrants' strategies. Some knowledges and practices remain *off the map*, in part because they are disqualified by the epistemic and political thresholds of the 'citizenship order', and in part as strategies of imperceptibility enacted by migrants to escape mechanisms of capture (Tsianos 2007). After all, the map traced out by migration policies is a counter-map, since it is based on those imperceptible knowledges, setting the boundaries to transform and translate (some of) these practices of movement into migrations that need be governed. In this sense, the theory of the autonomy of migration (Bojadzijev and Karakayali 2010; Mezzadra 2006, 2011c; Moulier-Boutang 2002; Rodriguez 1996) could be re-read along these lines: it refers less to the temporal primacy of migrations over the politics of controls than to the fundamental hijacking gesture through which migration policies act. It is just this off-the-map dimension that is of interest to Foucault when he refers to the subjugated knowledges that are objectified through human sciences and the production of knowledges on life (Foucault, 2003a). *Off-the-map* subjects are not merely silenced subjects to be heard, to be made to speak, or to be made visible, according to the logic of 'counter-acts'; instead, off-the-map subjects and practices introduce breaking points into the cartographic order, breaking open thresholds of perceptibility or interrupting mechanisms of capture. Moreover, insofar as these practices of struggle are *off the map*, they can take place without seeking formal political recognition, and in this sense they make destitute the legitimacy of power. In this light, analyses which centre on the appearance of uncounted and claimant subjects (Athanasiaou and Butler 2013; Isin 2002, 2012; Mouffe, 2005; Rancière, 2001, 2004a) risk taking for granted the boundaries of the political space of address and the desirability of such a (political) space, and inadvertently validate the functioning of power.

However, saying that knowledge about migrations needs to spy and capitalize on migrants' knowledge doesn't mean that governmental technology works only by reaction. In fact, along with that, the technologies of bordering and detention are characterized by their high degree of productivity (Anderson, Sharma, and Wright 2011) as well as by pre-emptive strategies and anticipatory logics (Amoore 2013). Borders not only cut (across) spaces but also produce differences in spaces—differences of status, differences of mobility, differences in the ways in which borders are enacted and crossed. To put it otherwise, the substantial productivity of borders and their transforming nature go along with an unceasing proliferation of borders. First of all, borders are traced through the current prolific discursive production of migration agencies which responds to the 'migratory disturbance', envisaging new spaces and times of governability (Hess 2012)—externalized protection, circular migration programs, mobility channels and humanitarian corridors. Nevertheless, this productivity of spaces should be better qualified: what is at stake is not only the proliferation of borders due to the tracing of spatial zones (zones of humanitarian protection, zones of detention and zones of free circulation) but also an uneven production of spaces. In fact, as Mezzadra and Neilson convincingly argue, addressing the 'equivocal character of borders', it is neither only a question of spatial scale nor of multiplication of borders: 'borders have not just proliferated. They are also undergoing complex transformations. . . . The multiple (legal and cultural, social and economic) components and institution of the border tend to tear apart from the magnetic line corresponding to the geopolitical line' (Mezzadra and Neilson 2013, 3). Migration policies instantiate *conditional spatialities*—that is, spaces that exist only for some categories of mobile people. The most pertinent example is the access to the European internal space of free mobility that third-country nationals can gain through the visa and Mobility Partnership centres, precisely in allowing certain categories of migrants to get it. Even if third countries establish Mobility Partnership with the European Union, migrants cannot circulate freely in the European space, but only in the member states that signed the agreement. Or if we think about the much promoted Euro-Mediterranean area of free exchange, it is quite evident that such a space, which does not have any geographical coordinates, really exists only for a very small percentage of the citizens of the southern shore of the Mediterranean: migrants coming by boat are not part of that economic and political picture. Thus, first of all, borders produce conditional and provisional spaces; but through migration policies and administrative measures of deportation and exclusion, they generate also exclusionary secure zones or spaces of citizenship that emerge as 'the result of the restrictions' (Karakayali and Rigo 2010). Consequently, the migration regime produces interruptions and introduces different forms and degrees of precarization. In fact, both collective migrant struggles and singular stories reveal that what characterizes governmental technologies for governing migrants is a substantial fragmentation of migrants' lives and journeys, imposing an uneven and unpredictable pace of (im)mobility. Such a fragmentation is produced through administrative measures or identification techniques, imposing indefinite stopovers or putting them in a 'bounce game' of bureaucratic hindrances, national boundaries, labour contracts and juridical decrees. The consequence is an indefinite lengthening of migrants' (interrupted) movements: thus, beyond blocking migrants' movements, what seems to emerge is the 'irregularity' of a protracted mobility in the twofold sense of the term 'irregular': both as a form of mobility 'illegalized' and as a production of discontinuity in the movements. At the same time, migrants' lives are differentiated through forms and degrees of economic and existential precarization, according to a variable range of mobility profiles which goes from economic migrants up to mobile person/non-migrants (Neilson and Rossiter 2008). The disposability of migrants' time is coupled with labour policies which push for a constant turnover of migrant labour force and hampering instead any persistence on the European soil.

CONDITIONAL SPACES TROUBLED BY TUNISIAN MIGRANTS

Tunisian migrants' spatial upheavals shook and destabilized this logic: they enacted their freedom of movement through collective departures and were indifferent to the conditional spaces of free mobility traced by migration policies. But they troubled the differentiated regime of spaces not because they moved illegally—this is the case with undocumented migrations, which are not revolutionary per se since illegalism is also part of migration governmentality. Rather, their disruptive force lay in the fact of exceeding the normal terms of illegalism itself, arriving suddenly in large numbers on the tiny island of Lampedusa, without demanding any protection but only to be released so they could move on. Moreover, they mocked the pace of mobility established by Europe, which after the Tunisian revolution paved the way for 'ordered' and selected channels of mobility to Europe to guarantee a 'smooth transition to democracy'. Tunisian migrants disturbed the uneven spatialities of free circulation because they did not come with the purpose of living in Europe or to find a job, as expected by migration policies: most of them, especially the youngest, left Tunisia just seizing the opportunity to visit Europe, and especially Paris, where they had Tunisian relatives and friends. To put it differently, their subversive practice and their 'scandal' consisted in inverting the direction of ordinary flows of bona fide travellers—European tourists going to North Africa—by actualizing their will to travel across Europe.

Ultimately, it is not even at the discursive level—through claims and demands—that they troubled the exclusive order of mobility, but through their effective enactment of the freedom of movement. They ultimately stepped out of the discursive register that can be easily recuperated by the state's narrative. Tunisian migrants not only staged radical statements. Nor did they simply make clear the contradictions of democracy—claiming that there cannot be democracy without freedom to move. They actualised or staged these contradictions not through the established channels of demanding formal recognition, but by engaging in practices of movement and 'spatial insistence' (Sossi 2013). Nevertheless, the question is not whether they effectively took part in the upheavals of 14 January or in the occupation of the Kasbah; rather, what is significant is that they asserted a relationship between the two movements, namely, the political uprising in Tunisia and their movements across the Mediterranean. Also in the case of nominal revolutions—

namely, when migrants' claims radically resignify and displace existing discourses and categories—these often reflect upheavals and forms of subjectivation that concern the materiality of people's presence and movements in spaces. Thus, the disorder that troubling mobilities produce is far from being narrowed to language. As Judith Revel poignantly puts it: 'disorder in subjective language . . . is undeniably effective, but that only represents one possible strategy inside an extremely thick set of resistances to objectification, hierarchization and exploitation' (Revel 2013b).

This coexistence of radical discursive claims and enacted practices of freedom not phrased in discourses leads us to a more general point on the relation between politics and language that concerns many migrant struggles. Indeed, if on the one hand, Tunisian migrants appropriated and radically restated the language of rights (mainly as the unconditioned right to move and stay in any place on Earth), on the other hand, the force of their upheaval relied on the not immediate translatability of their actions into rights claims addressed to institutions and governments. In other words, the discordant movements of Tunisian migrants were situated precisely on the edge between disruptive claims stating a universal and equalitarian right to move and non-discursive practices of freedom that (in part) slip the capture of the language of inclusion and episteme of citizenship. Actually, the difficulty in finding a space of address and an object of Tunisian migrants' challenge of the selected politics of mobility shows that in fact their practices of freedom neither claimed rights nor could be easily translated discursively.

The unconditioned right that Tunisian migrants actualized through their practices of freedom is actually an odd form of right, or better, a sort of 'non-juridical right' since, as Nicholas De Genova points out, it is not a question of migrants' rights but rather of the expression of migrant mobilization: 'that is to say they erupt from mobilities which cannot be fixed into place, categorized and regimented. They refer us to practices and processes of open-ended becoming, that actively produce and transform space' (De Genova 2010a, 115). If we look at these practices through the frame of rights, what gets lost is precisely the refusal to play within the space and the coordinates traced by the governmental migration map: in fact, more than claiming a right to freely move, Tunisian migrants made a space for themselves by taking it over, irrespective of its legitimacy, so sweeping away any conditionality and exclusionary access that the frame of rights necessarily instantiates (Honig 2006; Sossi 2012). Actually, the formula of the claim is in itself encapsulated within a regime of recognition that determined which practices were audible or visible; thus, the structure of the claim presupposes that subjects are expected to make demands for rights in accordance with the terms and conditions set by the 'governmental pact'.

However, the strategic importance that rights claims can play in migrant struggles should not be overlooked. Rather, it is a question of debunking the idea that rights are just a tool that can be used and adapted according to political goals and for supporting different struggles. Indeed, rights have historically functioned as the condition of a legitimate sovereign state, tracing the boundaries and the limits of power's exercise and, thus, as Derrida reminds us, rights are always an authorized force that de facto disqualifies as violent all the 'remnants', namely, practices and knowledges that cannot be integrated into the order of law (Derrida, 2006). Therefore, the battle over rights cannot but be played within the cartography of 'liberal' sovereignties, forcing some of its borders and making space for new political subjects. But as Derrida himself explains, drawing on Benjamin's analysis of the general strike, there could be claims and protests that, starting from within the space of the expected and governed social conflict actually point to the very limits of it, becoming in this way troubling movements. In other words, migrants' rights claims become a source of trouble when claims and statements that are formally phrased within the order of (legitimate) politics turn out to upset its grounds by affirming intransitive rights—like the right of everyone to trample everywhere on the earth—fundamentally untenable in that political space. That is to say, migrants' claims could eventually be disruptive movements when, I suggest, a destituent attitude—that 'empties' existing laws of their legitimacy—is coupled with a constituent move that also through a rights claim actually refreshes and rephrases completely the meaning of rights—conceived not as something to demand but as the reality corresponding to their practices of movement. Nevertheless, the decodification and the emptying of norms and rights do not translate into another discourse or order of political claims but in spatial practices that cannot be framed through the logic of rights.

Beyond the limits and the possibilities of a strategic use of rights there is also another fundamental issue that remains unaddressed in critical analyses and that concerns the juridification of life. Indeed, by translating migrants' spatial takeover into rights claims it necessarily entails tracing boundaries over lives and practices of movement, encoding them into accepted categories and limits of action. This is the reason why a struggle which hinges on rights claims could find itself in the slippery position of tracing the boundaries to the forms of subjectivity it fights for, recalling what Costas Douzinas defines as an incessant codification of life (Douzinas 2007).

From a theoretical and political point of view, it should be considered that an analysis centred on rights fundamentally fails to grasp the indistinct array of administrative measures and techniques of normalization that take place below, at the edges and beyond the law. In other words, despite the proliferation of legal categories defining mobility profiles, the mechanisms of capture and containment of migrants' lives step outside of and dodge the juridical sphere. In his course on the 'Abnormal' at the College de France in 1975, Foucault states that the field of normalization is produced through the *doubling* of juridical objects with something that stays at the edges of the law, representing its infra-liminar elements. It follows that the subject of rights is displaced by the figure of the delinquent, as its ethical and psychological doublet: thus, the purpose of the medical-juridical expertise becomes to retrace 'misdeeds that do not break the law, or faults that are not illegal' (Foucault 2003b, 19). Coming back to our present and to the domain of migration governmentality, it should be suggested that a similar but inverse doubling move is at stake. Indeed, the model of the subject of rights is likewise inappropriate for understanding the effective functioning of powers, since it is precisely below and at the margins of the judiciable that mechanisms of containment and exclusionary protection act. Something cannot but fall out of the law. And in this sense a doublet is at play also in this context. However, it is precisely the juridical subjectivity that seems to double 'irregular conducts'—but without producing a subject of rights. In fact, juridical profiles overlay disciplinary mechanisms—that shape risky subjects and troubling mobilities—translating them into intelligible and standardized categories. To put it in a nutshell, the heteronomy of the mechanisms that capture migrants' lives and their entanglement with different knowledges and techniques is coupled with a substantial non-homogeneity to the domain of law. The effective subjectivity that is postulated in the supposed neutral political subject-form—the subject of rights—leads us to look at autonomous migrations as movements that exceed the economic push/pull factors as well as their expectedness and usefulness into the regime of mobility, playing cautiously with the very notions of 'autonomy' and 'subjectivity'. In fact, a reading of the autonomy-of-migration standpoint that frames it in terms of autonomous subjectivities risks reinstantiating and stabilizing the order of political (in)visibility and representation that (some) practices of migration actually undermine. Indeed, the argument that I put forward as the issue of the autonomy of migration is not simply a question concerning the excess of migrants' subjectivity in relation to the economic mechanisms of capture. Nor is it the name for the ability of migrants to escape and resist migration controls or to struggle against mechanisms of exploitation. Rather, based on the idea of migration as a strugglefield, it refers precisely to the actualization of movements and practices that at least in part sidestep and escape the time, conditions and vocabulary of politics as space of citizenship—what I would call the 'citizenship episteme'. In fact, beyond forcing the mechanisms of capture to reassess the strategy and reframe the discourse on migration, strategies of migration that engage in more or less visible struggles often—as in the case of the Tunisian migrants—neither address nor fully respond to the political language of integration or recognition, or in any case displace the very meaning of it. Jointly, it is the name for a perspective gaze that looks at migrations as practices that force the politics of mobility to reinvent its strategies of control and containment. This allows us to fundamentally reverse the standpoint on the nexus migration-government, focusing on the supposed solidity of the migration regime as a set of mechanisms of capture. Finally, and maybe most importantly for the spatial perspective of this work, autonomous migration brings to the fore the spatial transformations and new cartographies that migrants put into place: migration movements do not only transform but also produce spaces—the spaces of their enacted geographies, essentially formed by the temporal borders they are confronted by—that only in part overlap with geopolitical boundaries.

Tunisian migrants effectively showed us the incongruity of the citizenship episteme for grasping the spatial upheavals they produced. Indeed, practices of migration were not simply a reaction to the liberation from dictatorship. Or better, while these were factors behind the decision to leave the country, they were intertwined with a fundamental drive to move, irrespective of migration policies and visa restrictions. Rather, for many of them the revolution mainly coincided with the actual possibility of fulfilling their wish of travelling across Europe for a shorter or longer period. If, on the one side, the revolution opened up concrete transformations within the country, at the same time it also paved the way for enacting the freedom to move out and experience other spaces. If it makes sense to speak of an 'excess' at play in migration movements, it is not due to an ability to get the best of the politics of control—since most of them have then been deported to Tunisia or blocked in Lampedusa—but rather in terms of the various desires and subjective drives related to the project of leaving the country that cannot be explained through liberation from the dictatorship. In fact, in the specific context of Tunisian migrations in the Mediterranean in the aftermath of the revolution, the discourse on autonomous migration could be aptly mobilized insofar as it refers to the *discordant practices of freedom* they enacted. By that I mean practices of freedom that reflect desires and projects that could be neither fully translated into nor easily recaptured by existing political claims and codes. Instead, regarding the crisis of the European border regime they momentarily produced, if we take into account the *Tunisian turmoil in Europe* on a more extended span of time, we see that most of the Tunisian migrants were deported or they came back to Tunisia because of the actual impossibility of living in France or Italy. In other words, the notion of autonomous migration should be handled with caution once we consider both the re-composition of powers and the consequences on migrants' lives in the long term.

Through their discordant practices of freedom they actualized a claim, more than stating it, which sounded like this: 'We, who actively took part in the revolution, are now here and are continuing it', stressing the hypocritical position of European governments, addressing especially France:

'You, who supported our democratic uprising and promoted the human rights rhetoric, are now chasing us away, the sons of the revolution'.[2] In the meantime, something has changed also in relation to the intolerable nature of power, namely, to the threshold of tolerance as regards the confinement of people's movements into wired places and, more widely, against their hampered mobility (Balibar and Brossat, 2011).

Moreover, many of those I encountered in Paris in 2011 were not at all meaning to remain long in France or to integrate into French society; it was most of all simply an opportunity to stay for a period in Europe, whether or not they found a job, an opportunity they never had before. Looking at the unexpected arrival of Tunisian migrants all coming in the span of a few months makes us interrogate the juridical and political transformations and the spatial turbulence that their presence generated on the European territory. New temporary spatialities emerged as the outcome of migrants' movements and, at the same time, as the effect of border enforcement at the level of European policies. As Federica Sossi persuasively argues, Tunisian migrants immediately had to confront techniques of bordering that were activated to respond to that 'sudden and effective upheaval of the space' (Sossi, 2012).

In order to see at work the discordant practices of freedom that Tunisian migrants put into place, I take a snapshot of the 'Tunisian revolution in Italy', 27 March 2011, Sicily: A group of Tunisian migrants clinging to the wire netting of the hosting centre of Mineo protest being detained like criminals. One of them screams, 'The world is not mine or yours, it neither belongs to Obama nor to Berlusconi, it belongs to everyone. So, if I want to breathe the air of Italy, I can do it; if I want to breathe the air of Canada, I can do it. No wire exists for me. I'm here not to steal or to rob; I'm here to breathe the air of freedom'.[3] What is stunning in these words, is, on the one hand, the insistence on an unconditioned freedom, and on the other hand, the inflection of freedom itself ultimately as freedom of movement, and at the same time as an unconditional right to stay everywhere, to trample the soil and to move on.

Moreover, such a claim reveals how social inequalities and class divisions are today reassembled through the geopolitical gap cutting across people on Earth who can move freely move with their passport, and those who require a permit. The air of freedom is precisely what he has been denied, first in Tunisia under the regime of Ben Alì—where the crime of emigration still exists—and then by European governments, who impeded his mobility and thwarted his desire to move on. This claim leads us to understand practices of migration as the strategies around which the migration regime (re)structures itself (Mezzadra, 2006). The fact that the Tunisian migrations that occurred in 2011 did not fit into the traditional schema of economic push-pull factors deeply undermines discourses and politics on poverty as the problem to be addressed: 'I don't want to be given any sandwich, I want to be left free to move away from Lampedusa'[4] —a Tunisian migrant firmly stated during the protests that took place in Lampedusa at the end of March 2011, when migrants who cried '*hurrya!*' ('freedom' in Arabic) asserted as neither negotiable nor deferrable their will to move. Yet both the Tunisian uprisings and migrations across the Mediterranean were narrated by the mainstream media as struggles for bread and for finding a job. These two issues were indeed at stake and it is important to acknowledge the existence and the legitimacy of these aspects in order not to fall into the error of detaching practices of freedom from any concrete concern, reiterating the gesture of seeing in events and practices what we are accustomed or willing to see. Nonetheless, economic reductionism cannot determine the complex strugglefields in which both migratory movements and revolutionary turmoil are situated, exceeding through their practices any demand that could be addressed to the existing border or power regime and even less answered by governmental actors. In this regard, Sandro Mezzadra draws attention to the connection between revolutionary uprisings and practices of migration: 'Why', Mezzadra asks, 'should that scream [*hurrya*] be bordered within institutionally defined spaces?' (Mezzadra 2011a, 116). The redefinition of the Schengen space and its principle of internal free mobility demanded by some European movements, along with the enforced blurring between politics of asylum and border controls, represent the two pillars upon which the European space of mobility has been reshaped to respond to migrants' spatial upheavals. Such spatial rearrangement signals less the enforcement of a European border regime than its uneven political geography and internal conflicts among member states. Thus, the emergence of conflicting spaces should be considered in all its complexity: on the one hand, migrants' crossings produced an upheaval of governed spaces, shaking the threshold of acceptability of power; on the other hand, the response given both by the European Union and by the Italian government didn't simply enforce existing techniques of bordering but rather paved the way for a deep reassessment of the European politics of mobility. In fact, at the European level the spatial rearrangement that followed the Tunisian turmoil consisted in a partial reconfiguration of the Schengen space[5] (Garelli 2013), while in Italy, through the invocation of the 'humanitarian tsunami', the Italian government enforced special and arbitrary measures of containment. Finally, in the name of a 'humanitarian crisis in North Africa' new actors and international agencies entered the machine of migration management and, broadly in the economy of development, produced new economic exceptional spaces of intervention.

TROUBLING MOBILITY AND THE APPARATUS OF SEQUESTRATION: REVERSING THE GAZE ON MIGRATION

In order to interrogate and unpack the unquestioned coupling 'migration government', or better, migration as a phenomenon to govern, I introduce Foucault and his notes on the spatial containment of 'irregular' mobility through the apparatus of sequestration and the hold over people's time. Indeed, as I will show in the chapter, although Foucault never directly deals with the issue of migration, his sideways approach to the theme brings to the fore and mobilizes a reversal of the gaze. In particular, I suggest that Foucault's reflections on the government of mobility in the 1973 course at the Collège de France *La société punitive* anticipate and dialogue at a distance with the theories of the autonomy of migration (De Genova 2010a; Mezzadra 2002; Mitropoulos, 2007; Moulier-Boutang 1998; Papadopoulos, Tsianos 2012), positing that the control over mobility in the eighteenth and nineteenth centuries functioned through apparatuses of sequestration of people's labour force and time. In particular, the 1973 course provides an analytical lens on the politics of mobility that enables us to see the government of migration as a set of strategies for taming and channelling the 'mobility disorders'. In this way, if we read Foucault's analyses on government in the light of the thesis advanced in *La société punitive* about mobility—namely, the idea that the control over mobility worked as a strategy for taming and containing 'dangerous' displacements—governmentality appears essentially as a strugglefield and as a strategic response to those troubling movements. A history of the present hinged on mobility control as a strategic response to *undisciplined conducts* allows a tracing back to the inception of a more or less structured politics of control over people's movements across Europe. Indeed, as Foucault shows, since the rise of the capitalist mode of production the political technology of government over lives has taken mobility as one of the main contentious factors. Nevertheless, in a similar way to all of Foucault's other genealogies, a history of the present does not mean to transpose the technology of control over people's mobility in the nineteenth century onto our contemporaneity to understand the current migration regime: rather, it is precisely through a series of dislocations and ongoing reinvestments that the migration strugglefield emerges as a conflicting site. But at the same time, despite the specific historical context in which Foucault grounds his analysis, what is at stake in his reflections is a substantial gaze reversal on migration and government that can resonate and be reactivated in contemporary spaces. In particular, from the spatial standpoint mobilized in this book, I suggest that Foucault's sideways gaze on mobility and migration through his articulated analysis on labour regimes and disciplinary power allows us to see borders and partitions that cannot be spatially located on a map. In fact, geopolitical frontiers and the borders traced by migration policies need to be entangled with multiple partitions and bordering processes that stem from other mechanisms of control.

One of the seminal reflections of the 1972 Course *La société punitive* for a critique of the government of migration that tries to decouple and displace the migration-government nexus is, as I said, the sideways approach to the topic of mobility. Indeed, the main subject of the book is neither human mobility nor the politics of control exercised on it but, rather, the emergence of what Foucault calls the 'punitive society', paving the way to the analysis of the birth of the prison later developed in *Discipline and Punish*. Or to put it better, at the core of his genealogy there is the emergence of a new economy of power—disciplinary power—that started in the sixteenth century and that has deeply modified the functioning of social institutions, well beyond the walls of the prison[6]: why and how since the eighteenth century has detention become the main punitive technique? In order to answer this question, Foucault is in fact forced to engage in a twofold move of widening and displacement: imprisonment as a punitive technique should be situated into a technology of government and control that, taking different forms and degrees, has been employed in different sites of capitalist society[7]; and the function of imprisoning and banishing dangerous subjects in order to eliminate their intolerable illegalisms has always been coupled with that of fixing people to the dispositives of production. Thus, the attention shifts from the prison as an institution to the economy of power relations in which the prison itself is located. This analytical move towards the economy of power is actually what makes Foucault engage with the production and the government of people's irregular mobility. In fact, the gesture of historically retracing the function of mobility control in the emerging capitalist society doesn't entail tracing a history of marginal subjects but rather points to the centrality of mobility as an object of government. In this way, more than a history that accounts for the 'silences' in the main narrative of capitalism, a genealogy of the government of 'turbulent' mobility works as an analytical lens for grasping the economy of power and its contested nature from a different angle. This methodological approach could be put in resonance with what Silvia Federici argues about the relevance of a history from the viewpoint of women, clarifying that 'women . . . signify not just a hidden history that needs to be made visible but a particular form of exploitation, and therefore a unique perspective from which to reconsider the history of capitalist relations' (Federici 2004, 13). In particular, by taking on mobility as a main concern of the government of conducts since the modern age, it becomes possible—in the case of Foucault via the analysis of the penal system—to see 'the nature of struggles that, in a society, take place around power' (Foucault 2013a, 14). Therefore, to analyze the techniques of containment and punishment through and within the economy of power that sustains them doesn't at all entail neutralizing troubling practices and disruptive movements into a narrow economic explanation. On the contrary, by looking at the regulation of mobility, interrogating the economy of power in which this happens and analyzing the forms of this control, Foucault brings to the surface the responsive nature of disciplining technologies: in the face of the emergence of new forms of mobility and of illegalisms, the multiple conduits of power have to rearrange themselves into new assemblages. Indeed, they are apparatuses of capture for taming and extracting value from those unauthorized movements. In this regard, it could be argued that the control of mobility is also an attempt to retake control of the excesses of illegalisms and irregularities.[8] Before coming back to the gaze's reversal on migration mobilized by Foucault, I will dwell upon the targets and the functions of this modern obsession that mobility is a phenomenon to govern.

THE CONTAINMENT OF FREEDOM AND THE GOVERNMENT OF IRREGULAR CONDUCTS THROUGH THE CONTROL OF MOBILITY

Why did human mobility start to be framed in the seventeenth century as a dangerous and troubling phenomenon to control? Is it movement in itself that is a source of risk, and what is the 'scandal' of free movements? It is noteworthy that the explanation given by Foucault correlates with his analysis of disciplinary power. Indeed, the question on the function of incarceration as a mechanism of punishment is displaced by Foucault towards the transformations of the economy of power during the rise of capitalism: the institution of the prison, as part of a broader political technology, aims at having a hold on people's lifetime. In the eighteenth century the refusal to work became an object of moral condemnation and, simultaneously, of juridical sanction: 'There is a fundamental identity between the fact of moving and the refusal of work' (Foucault 2013a, 49). It follows that it is not complicated to understand why Foucault identifies in the vagabond the paradigmatic character of the deviant subject, 'the matrix of irregular conducts'; or better, the vagabond becomes the criminal par excellence, the individual in which moral deprivation, social instability and resistance to produce coincide (Chamayou 2010). At that stage of the capitalist society, dislocation from the territory and uncontrolled mobility started to be considered a source of instability and trouble, since it indicated a substantial unreliability of the worker's commitment.

Then, following the historical steps marked in Foucault's analysis, we are confronted with a slightly different scene: in the nineteenth century, states and employers try to bridle the labour force of the working class through the regime of contract. Therefore, it is less the marginal subject or the vagabond whose movements are controlled and hampered than the industrial worker: 'the point of application is no longer wealth as an object of possible appropriation but the body of the worker as productive force' (Foucault 2013a, 192). The refusal to put one's own body to work represents the major source of risk for capitalism, preventing life from being synthetized in labour force. It is precisely at this stage that the norm intervenes to guarantee the transformation of life into a docile productive body through the acquisition of labour discipline as a habitude. The norm works as a multiple fixing mechanism, which binds physically and morally the individual to a social space, to the discipline of labour and to the apparatus of production. Indeed, by linking closely the emergence of disciplinary power and the rise of capitalism, both Marx and Foucault unfold the unnaturalness of labour, highlighting the necessary surplus generated by the norm—and formed by coercive measures and moralizing technologies—to put subjectivities to work. 'Labour is absolutely not man's concrete essence or man's existence in its concrete form. In order for men to be brought into labour, to be tied to labour, an operation is necessary, or a complex series of operations by which men are effectively—not analytically or synthetically—bound to the production apparatus for which they labour' (Foucault 1994g, 86). The norm operates less as an exclusionary measure partitioning between normal and abnormal than by tracing a social space in which bodies and subjectivities are translated into conducts whose social bond is not given by property but by the fixation to the productive apparatuses. Hence, the incessant striving to produce a labour force and make of it a natural quality of subjects is primarily a mechanism for destituting and hindering autonomous conducts that, as depraved or lazy behaviours, refuse to become labour force. In fact, as Paolo Virno stresses, the potentialities of labour force—as sum of the different human faculties—is the precondition of biopolitics (Virno 2002).

Labour force, Foucault specifies, does not indicate only individual bodies but also collective subjects. Indeed, the dangerousness of uncontrolled mobility and the slippery continuum with other forms of illegalism that it establishes depends on the collective forms it can take: 'modes of existence', as Foucault defines them, that try to resist, circumvent and subtract from the capitalization of life into productive force, generating 'a counter-collective that could menace the institution itself' (Foucault 2013a, 129). Ultimately, the menace relies on the constitutive ambivalent relation between labour and capital: 'labour within capitalist social relations is, in this sense, always simultaneously labour for capital and also against capital' (De Genova 2013c). In this way irregular mobility is prominently a practice of spatial flight from a control over life's time that tries to bind subjects to the mechanism of production. A very similar thesis has been advanced by Yann Moulier-Boutang, who in his seminal book *De l'esclavage au salariat* explains the rationale that is at the basis of mobility controls and migration policies both in Europe and in the colonies: 'The control of the flight of waged workers represents the most important element that presided over the birth, the attrition and the substitution of different forms of non-free labour, and to the origin of social protection as well as to the status of free wage labour' (Moulier-Boutang 2002, 17).

THE GOVERNMENT THROUGH NONGOVERNMENT AND THE FRAGMENTATION OF MIGRANTS' TIMES

Taking together *Discipline and Punish* and *La société punitive* allows us to see the twofold effect of incarceration and the disciplining power at large: the production of illegalism and the transformation of life into disciplined labour force. In this way, the thesis of *Discipline and Punish* cannot be read in relation to the government of the mob without considering the 'apparatuses of sequestration' acting on marginal subjects and later on the working class. However, as I stated initially, Foucault's history of the present is characterized by the attempt to find switching elements and moments, indicating how certain mechanisms of capture are re-inscribed in different political technologies in order to respond to refusals to work. This is the reason why the picture traced by Foucault cannot be applied as it is to the current political frame, since, for instance, the heterogeneity of the present regimes of labour corresponds to likewise manifold labour conditions, constrictions and strategies of resistance. Beyond that, I focus here on what could appear as two limit cases in which the economy of power shifts in part from its ordinary functioning: the economic transformations during a period of crisis and the government of refugees. Actually, the functioning of apparatuses of sequestration described by Foucault in relation to the eighteenth and nineteenth centuries cannot simply be extended to any form of mobility. And in this regard the government of refugees emerges as one of the most long-standing peculiar regimes: the creation in the twentieth century of the system of asylum has traced a series of partitions between different kinds of migration—based on the main watershed between economic migrants and refugees. As far as the politics of asylum is concerned, it becomes hard to fully adopt and apply the analytical grid of the valorization of life's time for transforming life into a productive labour force. Indeed, the blueprint of the productive subject to be disciplined and exploited turns out to be only in part tenable: unlike the detention of irregular economic migrants as an apparatus for filtering and decelerating the incoming of the migrant labour force (Mezzadra 2006; Papadopoulos, Stephenson, and Tsianos 2008), the functioning of refugee camps and of the international protection regime works by stranding, breaking and fragmenting people's lives. This produces an ongoing interruption and detour in regard to the fixation to a linear and constant productive function. While the mechanisms of capture over vagabonds or workers described by Foucault aim at capturing people's time and spatially fixing them to a certain location or function, in this case the main effect is a substantial fragmentation of temporal continuity, actions/projects and patterns. Thus, against this background we should interrogate the kind of governmental rationality upon which these fragmenting captures are exercised. It could be suggested that the sorting machine that separates refugees from economic migrants pertains rather to a logic of government of (migration) population that by tracing a moral geography of protection reallocates and distributes people through selective criteria, rejecting the majority of them from refugee status. A similar circumspection should be taken in relation to the current political and economic context. Indeed, especially in a period of economic crisis, the strategy of full employment is by a great deal far from the actual economic rationale: a generalized production of precariousness, based on differential degrees of partial and temporary employment, seems to better qualify the economic grasp on migrants' and non-migrants' lives. In this regard it should be questioned whether the general figure of illegalism is still nowadays that of dissipation: rather, the moral partition between the active and disciplined working citizen and the lazy 'irregular' subject doesn't work so smoothly anymore, due to the substantial inactivity in which both migrants and non-migrants are forced to live. This complicates and blurs the boundaries of the so-called migrant condition, and at the same time it partially undermines the rationale of putting and fixing people to work. However, this doesn't mean that in a time of economic crisis the capitalization and fixation of bodies to the apparatus of production through the discipline of work is no longer at stake: on the contrary, the point is how work itself has been recast through mechanisms of increasing and diffused precarization that render the migrant and non-migrant labour force constantly at disposal and, at the same time, stranded and unemployed for long periods. To put it in a nutshell, the 'migrantization' of life-time and labour-time increasingly also affects subjects who legally are non-migrants.

BEYOND EXCLUSION AND SMOOTH GOVERNMENTALITY: THE UNEVEN MACHINES OF CAPTURE OF MIGRANTS' LIVES

Despite these huge transformations and increasing complexities that are well encapsulated in the multiplied migration taxonomy, Foucault's gaze on the economy of power provides a distinctive analytical lens for grasping the politics of government over existences. In fact, Foucault takes his cues from a radical critique of the notion of exclusion as a fruitful paradigm for understanding power relations and the complex games between powers and resistances. However, through this move Foucault doesn't deny the existence of mechanisms of exclusion; rather, he points out that the notion of exclusion is fundamentally grounded in the field of 'juridical, political and moral representation' (Foucault 2013a, 7). Indeed, the category of exclusion induces a sort of double overshadowing: on the one hand, it conflates all marginal individuals into a negative and lacking subjectivity—the repressed, the excluded—in contraposition to the full citizen; on the other hand, it doesn't look at the effective functioning of power relations, bringing attention to the forms in which power is supposed to work or appears. In the same vein, in *The Will to Knowledge* Foucault explains that his critique of the repressive hypothesis aims 'less at showing it to be mistaken than at putting it back within a general economy of discourses' (Foucault 1998, 11). Such a perspective becomes particularly useful in the context of the government of migrations. Or better, it helps us to avoid any possible binary opposition between integration/exploitation of migrants in the productive system and rejection/exclusion to produce bare life. Starting from this set of considerations, challenging the inclusion/exclusion paradigm means shifting the attention from the productivity/unproductivity of the subject to the struggle-field of governmentality, namely, to the economy of powers as confronted with all practices of flight and resistance that force the strategies of capture to reassess their strategy. Following this, I suggest taking into account the diverse political techniques that underpin the government of different forms of mobility, not through the double code of inclusion/exclusion, but through detecting the points and the mechanisms upon which it applies—for instance, the government of migration population, the selection of labour force or finally what I would call the 'government through non-government'. This last expression encapsulates well a quite common way of operating in the field of migration management that makes us in part reformulate the Foucaultian matrix of government as framed by Governmentality Studies. Indeed, as far as migration as an object of government is concerned, an exclusive focus on governmental mentalities tends to overlook the always existing gaps between governmental texts and the effective drive of strategies of containment and capture. Rather, the image of a capillary and at the same time a homogeneous machine of governmentality crumbles as soon as we are confronted with the infra-liminar level of administrative techniques, bureaucratic conundrums, practices of detention, management of 'humanitarian emergencies' and migrants' flights. The 'text' of migration governmentality staged in EU documents and in the discourses of migration agencies is undercut by the plethora of regulative practices that sidestep the supposed standardized governmentality and in part effectively work through a substantial nongovernment. Governing through nongovernment doesn't signify loss of control or a more slackened grasp on migrants' lives. On the contrary, it indicates that the supposed exhaustive *prise en charge* of migrants' lives through techniques of surveillance and through standardized procedures is actually confronted and combined with practices of non-intervention and discharge that leave migrants stranded in their legal, spatial or existential impasses and forms of immobility. To put it differently, the fragmentation of migrants' patterns and life time is eminently produced through the misfiring of mechanisms that actually aim at hampering people's movements, but that, in principle, are presented as source of a smooth governmentality that should in an orderly way distribute, select and allocate people in space. Ultimately, the fragmentation and the indefinite stalemate of migrants' lives is a tangible effect of a way of governing through nongovernment. In this way, in place of a substantive definition of governmentality involving a strong commitment to rationalities of government, Foucault's definition of government—as 'an action upon an action, on existing actions or on those which may arise in the present or the future actions'—allows a more flexible and nuanced notion of migration government, one that does not conform to a model of absolute and seamless migration management (Foucault 1982, 789). At the same time, it allows taking into account modalities of regulation that work by leaving people or dropping them off but that also have concrete effects on the possibility of others' actions in the fragmentation of life's time that they produce.

Coming back to Foucault's displacement of the binary inclusion/exclusion, it could be suggested that, along with the implications illustrated above, it enables a fundamental gaze reversal that, as I will show, leads us to the theories of the autonomy of migration. In fact, a disengagement from the code of exclusion first of all engenders a radical turn of the analytical standpoint, since reading the government of mobility through the lens of inclusion/exclusion means assuming the order of citizenship as the given space against which any form of 'irregular' mobility is confronted. The refusal to adopt such a citizen's gaze allows looking at practices of mobility as movements that induce what Nicholas De Genova calls 'reactionary formations', namely, strategies and techniques of capture that try to contain, filter and tame those troubling mobilities (De Genova 2013c). Thus, to speak of practices of movement in terms of migration and government is already a form of translation and a way to encode 'irregular' mobility into the order of citizenship, as an abnormal and disturbing factor, a deviation from it. In other words, the *unbearableness of irregular mobility* relies precisely on the adoption of the State and citizen standpoint that makes us see practices of movement as instabilities to the channelling or capitalizing of labour force. Instead, as Dimitris Papadopoulos points out, reflecting on the practices of flight that characterized the epoch of the rise of capitalism, 'the singularities that composed the escaping, wandering mob were very far from the humanist individual emerging at the same moment across Europe . . . collectivities that defined the core of radical struggles for emancipation' (Papadopoulos, 2010). And, recalling closely the analysis of Moulier-Boutang on the emergence of the regime of wage labour, he continues by saying that 'many of the scattered, disorganised, ephemeral, insurgent movements of people exiting feudal labour in so many different locales and geographies, continents and seas were not to enter into the capitalist humanist regime of the labour market but to escape into a journey that allowed them to create common worlds' (Papadopoulos 2010). Therefore, it is now quite evident that such a gaze reversal on migration provides also a similarly discordant gaze on the history of capital-labour relations. Mobility turns out to be at the very junction of the strugglefield around freedom and property: if the government of mobility should be read as a way of controlling productivity, this latter must in turn be understood as a containment of freedom. In fact, as Foucault also indicated, presenting the control over mobility as a way for governing conducts and bodies against their practices of freedom and flight, Moulier-Boutang underlines that 'freedom is essentially the freedom to escape the land of the landlord and, at the same time and immediately, the wage labour' (Moulier-Boutang, 272). Foucault's insight on the raising of controls over mobility in capitalist society and theories on autonomous migration enable a challenge also to the current securitarian rationality that sustains the legitimacy of the government of migration: in fact, if migration controls are conceived as a strategy for containing freedom, it follows that the question on the failure of migration policies in reducing the number of migrants is wrongly framed. The illegalization of certain forms of mobility and techniques of surveillance 'engender an acceleration of the processes of mobility' and 'in order to escape sanctions [populations] tend to conduct an irregular life and fall into nomadism' (Moulier-Boutang 2002, 195). Hence, beyond any government of borders and numbers, we see that the government of mobility is first and foremost a technology for managing populations and conducts through a containment of freedom.

From this perspective it becomes clear why, as Foucault stresses, the criminal and the vagabond collapse into the figure of the dangerous individual: the uncontrolled mobile subject, who tries to escape the norms of society and forces the regime of productivity to negotiate the mechanisms of exploitation and invent new strategies of control. Beyond that, 'irregular' mobility has been seen by governments and states since the late eighteenth century as the troubling element of the existing political ontology: indeed, although governmental actors frame migration in terms of mere geographic dislocation, practices of movement as resistances to societal normativity are also transformative forces producing and altering spaces. 'Migration is not the evacuation of a place and the occupation of a different one, it is the making and remaking of one's own life on the scenery of the world. . . . Its target is not relocation but the active transformation of social space' (Papadopoulos and Tsianos 2012).

REPOSITIONING THE AUTONOMY IN MIGRATION IN THE LIGHT OF MIGRANTS' PRACTICES OF FREEDOM

However, what cannot pass unnoticed is that all these irregular mobilities are defined by Foucault as practices of flight, resistance and refusal. This is important in order not to overrate or misunderstand the notion of the autonomy of migration as framed by the above mentioned authors, which also emerges indirectly in Foucault's analyses on the punitive society. Indeed, the term 'autonomy' does not refer, I suggest, to a supposed autonomous subject able to sidestep and exceed the mazes of power, norms and economic exploitation through a free and insurgent act of migration. It goes without saying that such a description would contribute, on the one hand, to the fostering of an image of migrants as heroic or revolutionary subjects, and on the other, to overlooking the impact of both geopolitical and non-geopolitical borders on migrants' lives and the anticipatory moves and strategies of capture that today characterize migration governmentality. Moreover, since regulated human mobility is one of the main conditions for the existence of capitalism, migration as such is far from being a revolutionary practice in itself: as Mezzadra remarkably puts it, 'There is no capitalism without migration, one could say, with the regime that attempts to control or tame the mobility of labour playing a strategic role in the constitution of capitalism and class relations' (Mezzadra 2011c). From a Foucaultian point of view it is important to find the specific moments and contexts in which some practices of migration have effectively troubled the order of the politics of mobility. Therefore, the vantage point of the autonomy of migration as introduced by Foucault in his account of the disciplining of mobility is on the edge between a historical and punctual analysis and a general statement: in practice, only sometimes does migration really interrupt or disrupt the order of mobility; but it remains that the very existence of a migration regime is itself a responsive strategy to practices of freedom that try to escape economic or social captures.

Instead, practices of resistance and acts of refusal as practices of migration are always practices of freedom to escape and struggle with the economic and social dispositives in place. This is the reason why more than reading it in temporal terms—practices of mobility anticipate power's mechanisms—the perspective of the autonomy of migration should be posited as a vantage point through which to look at the functioning and transformations of the migration regime: migration policies, techniques of surveillance and strategies of patrolling represent the most visible and tangible elements of the multifaceted frantic attempt to bridle mobility and capitalize it. And the excess that some migrations enact in relation to the given relations of force should not be seen as an ontological primacy but rather as a practice of freedom against the existing machines of capture and the moral geographies governing migration population. Ultimately, the political and troubling character of (some) practices of migration emerges only as far as migration is taken as a specific kind of social relation and in its imbrication in a certain strugglefield. Thus, in Foucault's analysis it is less the migrant labour force as such that is highlighted as a troubling factor than the strategies of resistance and the practices of freedom enacted by people refusing government of conducts and bodies. It follows that by reading practices of movement not exclusively as a subtraction from control but, in a broader way, as forms of dissidence and resistance to the moral, economic and political government over lives allows going beyond the juridical paradigm/blueprint and finally 'cutting off the head of the king'. The critique of the binary code exclusion/inclusion goes hand in hand with the displacement of the centrality of law with its sorting function—tracing the boundaries of irregularity: 'the struggle against coercion is not the same as overcoming the prohibition. . . . To enact a transgression means making the law unreal and powerless, for a moment, in a certain place and for one person; becoming dissident is to attack this coercion' (Foucault 2013a, 116) and, we can add, to enact a practice of freedom that is discordant, non-representable, into the existing cartography of the political.

Looking at (autonomous) migration through a counter-mapping perspective means engaging in the production of not so much a counter-narrative on migration as another map and discourse which displaces the migration-government nexus as an unquestioned blueprint to speak of people's movements.

THE PARADIGM OF THE ACTIVE MIGRANT-CITIZEN: A CRITICAL ACCOUNT

If we confront a gaze that takes migration as a vantage point with the literature on acts of citizenship and theories of radical democracy, what characterizes the latter is the 'script of interruption' through which the relation between political space and subjectivities is framed. By 'script of interruption' I mean the bursting into the scene of the political by 'claimant subjects' who perform 'an act that does not exist or an act that already exists but that is enacted by a political subject who does not exist in the eyes of the law. Presences that defy where an act can be staged' (Isin 2012). From such a perspective, politics is designated as the action which transforms the space into a space for the appearance of the subject (Rancière 2001). According to this discourse, the moment of the interruption coincides with the emergence of political subjectivities. Thus, subjects who are at the edges of representative politics are supposed to lay a claim to be counted into the order of citizenship, stretching boundaries and norms: migrants become the central figures of these analyses due to the 'subject position' they embody. These readings have played an eminent function in reversing traditional political postures, by drawing the attention to the ways in which the 'inside'—the space of citizenship—is defeated and rearticulated by the 'outside' through the acts of those subjects who are excluded from that space.

The *border interruptions* produced by Tunisian migrants in the time of the Arab revolutions were predicated neither on the possibility of transforming the regime of borders and visibility nor on a claim to be included in the space of citizenship. In fact, as Ranabir Samaddar remarks, the emergence of a political subject which is 'out of place' or 'unexpected' in a given political space (as migrants) 'exceeds the rule of politics' (Samaddar 2009): it always comes out as a supplement regarding that space, and it is 'fundamentally a matter of non-correspondence with the dominant reality. Political subject exceeds rules of politics' (Samaddar 2009, 16). This is why the term 'citizen' cannot encapsulate the huge variety of material processes and practices through which (and against which) subjectivities are produced. Instead, the script of the 'scene of answerability' that underlies those analyses on citizenship and migrants' subjectivity leaves a huge array of migrant struggles under the threshold of perceptibility, relegating them as pre-political languages often failing the issue of political organization as well as the production of a common space of address. Analysing the revolts in the French banlieus in 2005, Rancière contends that 'they are political because they are the response to a situation of inequality. But the response has not really been political. . . . Politics requires more than these forms of awareness and of collective revolt. It requires the constitution of a polemical common space, a space of address' (Rancière, 2005). The Tunisian migrants—who most of the time are undocumented migrants who interrupt for some moments the functioning of borders—neither engage in making explicit claims nor performing creative acts: in this sense, the scheme of 'agency' is grounded on a too 'exigent' model of subjectivity, tracing out exclusionary borders of what can be defined a political practice and what acts produce political subjectivity. This is particularly evident in the case of Citizenship Studies, in which migrants' strategies are generally framed in terms of agency (Coutin 2011; McNevin 2006; Nyers 2006, 2011). This literature has contributed to reformulating the theory of the autonomy of migrations but stripping it of the conflicting dimension upon which this latter is predicated. Instead of taking migrants as the new active subjects who could contribute to reinvigorating the institution of citizenship which is today in crisis, I suggest drawing attention to the disaggregation and crises of citizenship that stem from migrants' slippery position. Indeed, also those who are formally inside the space of citizenship experience the overlapping of discrepant statuses and conditions that immediately detach the formal acquisition of citizenship from its effective enjoyment.

To develop the implications of the translation of autonomous migration to migrants' agency, I ground on the binary divisions that characterize critical Migration Studies and on the individualism that underlies the category of agency. Migrations are usually depicted as phenomena taking place in between an inside-outside political move: migrants are basically those who try to become part of a certain space or that stretch the borders of citizenship and belonging. By binary divisions I mean also the way of pigeonholing migrants' practices of resistance according to clear profiles of subjectivities fixed in advance: claimant subjects, victims, vulnerable subjects, agent subjects. While the theory of the autonomy of migrations stresses the collective dimension of migrants' strategies, framing them in terms of 'movements', the focus on agency instead brings attention to individual abilities and will in challenging the border of the political, reproducing an image of subjectivity grounded on the liberal paradigm. As Papadopoulos and Tsianos point out, drawing on the autonomy of migration framework, 'the concept of migration does not mean flattening out their differences; rather, it attempts to articulate their commonalities which stem from all these different struggles for movement' (Papadopoulos and Tsianos 2013, 185).

In other words, two different ideas of autonomy sustain migration analyses on agency and the theory of the autonomy of migration: while the latter conceives of autonomy within the frame of collective movements, drawing on the workerist tradition, the former refers to agency as depicting autonomous subjects unsettling and addressing the scene of visibility. Secondly, the emphasis on agency and on symbolic gestures tends to overlook the material conditions in which many undocumented migrants live, and thus the deadlocks migrants can come across in enacting public protests. The third element of criticism concerns the temporality of struggles: in fact, agency is focused on the moment of the disruptive act. Shifting from the moment of border crossing towards the 'migrant condition'—as a concrete practice of living that (undocumented) migrants experience daily—makes it possible to locate migrants' practices on a temporal dimension usually dismissed by activists. In fact, once a given discursive frame or a political space is cracked by migrants' presence, what gets lost are the consequences of these acts on migrants themselves, given that deportability is the primary weapon that states have (De Genova 2010b). Instead, by assuming agency as the prism to codify migrants' struggles, one risks reiterating the theoretical gesture that designates as 'disruptive' and 'political' those practices which can easily be codified or understood through existing categories. To dwell upon what comes before and after the disruptive moment means to pay attention to the consequences of producing interruptions for those who are not part of the 'citizen contract': the detainability and deportability of undocumented migrants are the conditions upon which the fragmented temporality of migrants' practices is predicated. Thus, the coupling of transformation and interruption can take place only when practices endure over time: the challenge of the subject position that migrants are expected to fill does not necessarily correspond to the production of a different order. Quite to the contrary, the condition of being spoken of and labelled by power, becoming visible only in a clash with it—in its discourses or through its administrative techniques of identification—makes it difficult for destituent practices to transform into constituent movements. Also, the possibility for subjects to depart from migration categories is a quite wearying task. After all, the vanishing of migrants from the public scene relates to the lack of traces that their presence leaves on the ground: in other words, how to account for subjects who can neither emerge on the 'scene of address' nor be narrated except through encountering power? In this regard, Foucault's text *Lives of Infamous Men* stages this point very clearly: 'In order for some of them to reach us, a beam of light had to illuminate them, for a moment at least. A light coming from elsewhere. What snatched them from the darkness in which they could, perhaps should, have remained was the encounter with power' (Foucault 1994c, 161). By narrowing politics to the moment of rupture one fails to take into account the intense and more invisible activity of norms in shaping subjects. Foucault's analyses on the production of individuals through the working of norms and knowledges highlights that the focus on subjects making interruptions of borders cannot be detached from a questioning of the costs and conditions of becoming a subject. In other words, the fact of being identified as an undocumented migrant embodies very concrete conditions of (in)visibility and exploitability which shape the daily strategies of existence (Sossi, 2007).

THE EXCLUSIONARY SPACE OF ADDRESS

In the face of analyses that frame migrants' struggles as claims to integration within the political space of citizenship, some scholars have countered that citizenship is constantly defeated by migrants who upset the boundaries and the conditions of being political (Isin 2002, 2006; Nyers, 2003, 2011; Rygel, 2011). Secondly, they put into motion and stretch the notion of citizenship, assuming it not only as a juridical status but as the result of acts of claims. However, what remains unquestioned in these analyses is the validity of the paradigm of citizenship for speaking of a multiplicity of migrant struggles that actually neither claim citizen status nor demand to be included in the boundaries of civil society. And, ultimately, they postulate a model of political subjectivity and of political movement shaped according to the coordinates of Western activism. Indeed, as Angela Mitropolous remarks, migrants are ultimately seen as subjects who need to struggle to become political and 'talking' subjects: 'political subjectivity is invoked on condition of assuming the perspective of the State', and migration is assumed as 'implying the absence of political decision and action' (Mitropoulos 2007, 11). In this regard, let's turn the attention to the presence of thousands of Tunisian migrants in the streets of Lampedusa. March 2011: The migrants' claim 'we don't want to stay here—in Italy—we want to move away and go to Europe'[9] resounded on the overcrowded island in response to the cameras and journalists that were there to foster the 'spectacle of the border' (Cuttitta 2012; De Genova 2011). After all, as sons of the Tunisian revolution they did not come here to find out what democracy is, but rather often presented themselves as those who made their own revolution for democracy in Tunisia, displacing the common imaginary of Europe as the most desirable space and of migrants coming from destitute and oppressed lands to come in contact with the 'advanced democratic societies' (Rose 1996, 1999).

This episode raises a question about the possibility and the opportunity to locate in advance the space of address of migrant struggles (Rancière, 1995). In fact, by designating what struggles stand for, we definitively trace the map and the borders of the space of their political movement, weakening their disruptive force. Encoding practices into a pre-established language of claims and into a given space of address means to translate migrants' practices into the frame of recognition: the 'scandal of democracy' cannot be grasped if the interruptions of/at the borders are framed in advance as democratically oriented movements and demands. Indeed, many migrant struggles are not struggles because they challenge the discourse or the mechanisms of power or because they address political institutions, but rather insofar as they are involved in complex strugglefields, namely, economic, administrative and discursive borders which migrants have to confront and by which they are shaped (Mezzadra 2007, 2011c). This calls into question the 'outside' position usually attributed in critical literature to undocumented migrants who try to disrupt and stretch the borders of the 'inside'—the political space of citizenship—as if they could be outside of any economic-political relation to power. Instead, the contested political space that both Migration Studies and Citizenship Studies focus on is situated in complex economic relations that by now play at a transnational level, and it intertwines with many kinds of spatialities—for instance, different spatial economies. Hence, it becomes problematic to pose an 'outside' where migrants would be before being 'captured' or integrated. After all, the degrees of 'inside' and 'outside' exceed the official demarcation of a political space of citizenship: as Partha Chatterjee shows, there are citizens who are de facto excluded by the realm of civil society and that conversely are subject to the 'grasp' of governmental power (Chatterjee, 2005). Before and simultaneously with struggles that crack the functioning and the legitimacy of the borders of the 'political' space of citizenship, migrants (as well as non-migrants) live within the meshes of multiple powers, among which the economic mechanisms of extraction of surplus-value are certainly the most confining (Chignola and Mezzadra 2012). And Foucault's analytic of governmentality allows challenging any sharp inside/outside spatial divide, not only displacing and multiplying the sites of power but also positing an inextricable interweaving between the economics and techniques of government. In fact, in Foucault's account of governmentality, economy is posited as the main knowledge through which governmental power works; its emergence as a discipline corresponds to the affirmation of modern governmentality and the functioning of the mechanisms of security. At the same time, the economy that Foucault takes into account concerns power relations at large, namely, the economy of power. In other words, economy stands also for the mechanisms and rationalities that determine the coherence of a technology of power. What is in question is not so much the articulation between the political and the economic but their mutual constitution and finally the possibility of reading the former through the lens of the latter. In fact, if it is true that Foucault stresses the autonomy and the specificity of power relations (Foucault, 1982, 1994b), the economy of power that he sees intermeshed and constitutive of the political is concerned with the practical governmental reason that regulates bodies and phenomena—'the government of men is a practice . . . that fixes the definition and respective positions of the governed and governors facing each other and in relation to each' (Foucault, 2009, 12).

Therefore, against the supposed exclusionary 'pure' political space, a critical focus on migrations suggests looking at the articulation and multiplication of different spatialities, challenging the idea of spaces and subjects outside the political. The very concept of space need not only to be pluralized but also frayed in its supposed homogeneity, especially regarding the differential access to mobility, which basically depends on how the migrant condition impacts differently on people's lives. From this standpoint, migrations impose on us the need to shift the spatial gaze from the regime of visibility/invisibility and circulation towards a comprehensive investigation of the conditions at which flows and visibility are produced, highlighting 'the inequalities of property and the process of labour exploitation' (Tsing 1997, 336). In other words, looking only at the political cartography traced by the liberal discourse on mobility, 'we lose sight of the material components through which sites are constructed and from which convincing claims about scale units and scales could be made. . . . A focus on circulation shows us the movement of people, things and ideas, but it does not show us how this movement depends on defining tracks and grounds, scales or units of agency. Flow itself always involves making terrain' (Tsing 1997, 338). Therefore, under what conditions and through what processes does a certain spatial configuration come to work as a baseline that posits the thresholds of perceptibility of practices and events? In fact, the critical account on the pitfall of the governmental grid (see chapter 1) leads us to a similar reflection on freedom. By framing Tunisian migrants' movements as practices of freedom we cannot avoid questioning the very notion of freedom and, in particular, the transversal striving for freedom of movement. The image of a space of (free) circulation is essentially misleading since, as Marx has extensively demonstrated, it conceals the mechanisms—of exploitation and exclusion—upon which it is grounded (Marx 1993). In contrast to this, Tunisian migrants staged the unbearable price of free circulation encapsulated in the mechanism of the visa system by simply enacting the freedom to travel as it is allowed to 'mobile people', and at the same time risking their lives at sea.

In his account of governmental reason and neoliberalism, Foucault addresses precisely the economy of power in which freedom, and in particular freedom of circulation, is situated and predicated, and jointly the kind of subjectivity produced and regulated by that political technology. Consequently, instead of positing a space—the space of free mobility, or the space of citizenship—as a bounded unit, and instead of assuming a subject who is supposed to claim recognition, one could look at strategies of migration investigating the material processes through which, from time to time, some people are labelled as migrant-subjects. Ultimately, the emergence of processes of subjectivation and production of subjectivity are at the core of critical analyses that aim at probing the transformations of power. Thus, in order to understand how the rationale of migration governmentality has changed over time, the issue of subjectivity—who is labelled as migrant and how migrants crack and exceed mechanisms of capture—cannot be left aside. In fact, far from being only the name of the liberal diagram of power, governmentality refers to the inextricable intertwining between production of subjectivity and power's regulation: 'talking about governmentality I refer the whole array of practices through which it's possible to constitute, define, organize and play with strategies that individuals, within the scope of their freedom, act on one another and on themselves' (Foucault 1996a, 448). And placing at the core the technologies through which people act upon themselves and on others, it makes it possible to come to grips with subjectivity beyond the subject of right, since 'governmentality, I believe, enables us to account for the freedom of the subject and its relations with others; that is, what constitutes the matters of ethics itself' (Foucault 1996a, 448).

NOTES

1. Interview with the Guardia di Finanza, Lampedusa, 30th January 2014.
2. Interview with Tunisian migrants in Paris, November 2011.
3. http://www.storiemigranti.org/spip.php?article922
4. Tunisian migrant arrived in Lampedusa, February 2011.

5. The first measure was taken by France, which, in April 2011, closed the border with Italy, refusing Tunisian migrants who received the humanitarian temporary protection visa to enter the country, breaching the Schengen normative framework according to which migrants with a permit obtained in a member state could circulate in the European space for three months. Then, in May 2011, both Denmark and the Netherlands declared that for 'security reasons' they would reactivate internal border controls. In September 2011 the European Commission proposed a modification to the Schengen border code stating that 'the reintroduction of controls at internal borders should be based on a decision proposed and adopted by the Commission' but 'for unforeseeable events, Member States retain the possibility to unilaterally reintroduce border control at internal borders, if immediate action is needed' (COM [2011] 560 final).

6. The functional continuum between the prison and other social institutions of 'sequestration', as Foucault puts it, is given also by the techniques of control on and through the time. In fact, the form prison and the form salary share the sequestration of people's time. 'As one gives a salary which corresponds to the time of labour, conversely one takes the time of freedom as the price to pay for an infraction. . . . What emerges through these two forms is the introduction of time into the system of capitalist power and into the penal system' (Foucault 2013a, 72–73).

7. 'The penitentiary is actually a phenomenon that is much larger than the imprisonment and it is a general dimension of all social controls' (Foucault 2013a, 104).

8. The excess is mainly determined in relation to the economic necessity (of labour force) and to the political function of illegalism: when, for economic or political reasons, certain forms of illegalism become counterproductive, they start to be sanctioned and banished, and they are framed as immoral conducts. While tolerated (and necessary) illegalisms are only the object of a juridical sanction that is then not enforced—they are tolerated—'excessive' illegalisms are instead entrapped in a moral-juridical double that presents them as dangerous and squandering conducts. But in both cases, as Foucault stresses, what is at stake is a quite anomalous juridical sanction: actually these irregularities are not infractions, since 'given the necessary freedom of labour markets, it's impossible to organize the juridical system in a way that [all these irregularities] can become infractions. Therefore this illegalism spreads at an infra-legal level' (Foucault 2013a, 196).

9. This was the common Tunisian migrants' claim when they were blocked on the island of Lampedusa.

THREE

'Which Europe?'

Migrants' Uneven Geographies and Counter-Mapping at the Limits of Representation

The Arab uprisings have been received and welcomed on the northern shore of the Mediterranean as the forerunner of new political forms of protest and re-significations of the public space. The occupation of the Kasbah in Tunis and of Tahrir Square in Cairo seemed to resonate strongly with movements of protest in reaction to the global economic crisis and against government politics of austerity. The simultaneity of the Occupy movements and the Indignados with the events of the Arab uprisings has clearly fostered the easy connections posited by many critical analysts, activists and scholars: 'From Tahrir Square to Oakland', 'From Cairo to Wall Street. Voices from the Global Spring', 'Egypt Supports Wisconsin', 'Turn Wall Street into Tahrir Square'. An inverse and double contamination seemed to be envisaged in all these analyses: the 'not yet' of democracy going south, and new fresh practices of political participation migrating from the southern Mediterranean countries northward. It is beyond the scope of this work to make a comparative analysis of the European and American Occupy protest movements and the revolutionary uprisings in the Arab region, or to investigate their mutual influences. Rather, and without denying this recognized nexus, I will focus here on the spatial upheaval triggered by the Arab uprisings on the northern shore of the Mediterranean and, in particular, on what could be named the *Tunisian upheaval in Europe*. The term *bank effect* refers here to the reverberations and the repercussions that migrations and the Arab uprisings generated on the northern shore of the Mediterranean.

However, instead of dwelling upon the temporary interruptions of the mechanisms of capture enacted by Tunisian migrants at the time of the Arab uprisings, I turn the attention to the spatial disruptions and reshaping that their turmoil triggered. In fact, the choice here is to shift away from the temporality of the event, not simply conflating spatial upheavals with punctual interruptions: in the face of the elusiveness of migrants' geographies and of the temporary 'sabotages' of the governmental migration map, with its apparatuses of capture, the issue becomes one of looking at whether and how 'another map' has been finally traced by all these movements.

Nevertheless, it doesn't follow that escapes, practices of migration and strategies of resistance are coextensive and isomorphic to power; rather, I focus mainly on the unsettling unexpected nature of resistances that try at times to dodge and at times to reverse the political technology aimed at capturing and capitalizing people's movements. In this account, while we certainly need to emphasize acts of refusal and flights from the mesh of power, at the same time I shall make room for the asymmetrical dimension of resistances conceiving that dimension precisely as the capacity for transgressing and redefining meanings and uses of spaces along with the invention of forms of struggle and strategies of survival.

By mobilizing the technique of snapshots, I turn attention to the production of new spatialities resulting from these practices of migration. Such spatialities are framed and enacted by political technologies through economic or juridical measures—measures to respond, catch, (re)frame, neutralize and profit from revolutionary events; but also such spatialities are subverted and produced by migrants crossing borders. The advantage of such a perspective is, as William Walters observes, that 'the zonal allows us to map irregular spaces that are neither national nor global and it avoids the teleological assumption that developments in migration control have an inevitable direction or end point' (Walters 2011b, 53–54). Such a spatial gaze, I contend, could work as a litmus paper to scrutinize the subtle dynamics and the instabilities between power over migrants' lives and the power of migrants themselves. In fact, while political technology continuously plays in space and by making spaces in order to govern people's movements—creating special emergency areas, devising advanced techniques of bordering or externalizing frontiers—at the same time, migrants' practices not only can trouble those channelled spaces of mobility but also put into place different modalities of enacting and persisting in space. The reorganization of the border regime in the face of migrants' spatial upheavals basically takes place through two main strategies: the *strategies of detection*—improving surveillance systems and monitoring techniques that displace the border before and beyond the geopolitical line—and the *strategies of b/ordering*—techniques and knowledges through which migrants' conducts are regulated in their spatial persistence and in their mobility.

The goal is not to bring out possible connections with events, practices and movements currently at stake on the northern shore but, on the contrary, to show how migrants' spatial upheavals have multiplied—triggering a sort of domino effect—and, most importantly, have engendered political and spatial transformations, unsettling the tempos and the conditions of the politics of mobility. But I tend to resist designating migrants' practices as 'political movements', including in the specific case of Tunisian migrations during the Arab Spring. Indeed, the risk is that we overshadow the specificity of practices of migration and of struggling as a migrant, encapsulating all under the overcharged category of 'political movement' and re-establishing the model of a collective homogenous subject form. At issue in these three snapshots are the instabilities produced at different levels by migrants' spatial upheavals and discordant practices of freedom that do not fit into the script of the insurgent democracy or of majority politics: they trouble the isomorphism of the surface that makes different political experiences easily connect and influence each other.

These snapshots aim to bring to the fore the coexistence of different spatialities with their nature of *emergent-emergency spaces*, meaning by that the production of new political practices as correlative of, at the same time, a temporal and a temporary dimension. 'Emergent' alludes to a temporal aspect, namely, to spaces that arise suddenly as the outcome of economic spatial constructions and juridical decrees or of strategic struggles and spatial practices. Turning the gaze to these spatialities, this chapter highlights the production both of anomalous spaces of governmentality and of spatial overturnings which challenge representability and rights as the main axes of the political domain. 'Emergency' refers to the temporary dimension carved out and vested in these spaces as political responses to a disruptive and troubling array of events, subjects and transformations framed as threats to the social and spatial order. Bordering processes and migrants' strategies do not take place on a flat and smooth surface: rather, they often give rise to spatialities that work as a kind of *training-anomalous-terrains* of political experimentation that then might spread to other domains or spaces, especially as circulation of political languages. The second related point considers that we should bypass the form of the 'camp' as the exceptional space par excellence (Agamben 1998) to understand what is at stake in the power over/of migrants lives; rather, the spatialities I deal with are characterized at once by a specific economy of spatial practices and by a movement which constantly encroaches on the outside.

Mapping the effects of the crisis in Europe, charting the new composition of urban spaces and tracing the cartography of new social movements, one of the most widespread analytical gestures today consists in mobilizing visual metaphors and devices related to the field of mapping and cartography. The 'cartographic anxiety' (Gregory 1994) of making everything visible and representable which is constitutive mapping, seems to currently pervade the production of knowledge in the domain of social, political and human sciences. The visualization of political and spatial transformations on a map, and the attempt to provide a spatial diagram that deploys and orients all phenomena and subjects in space, confront us with a politics of visibility that encounters specific limits and political implications when addressing practices of (unauthorized) migration. Against this background, this chapter tries to unpack and come to grips with the theoretical and political conundrums of mapping migration—both as a cartographic practice and as a form of narrative; it mobilizes a counter-mapping analytics which works at the very limits of representation.

Three migrant struggles—the *Lampedusa in Hamburg* collective, *Syrians stopped in Calais* and Eritrean migrants in Lampedusa refusing to give their fingerprints—are the snapshots that will be taken here, time-spaces of migrant struggles and of migrants' presence in spaces which have troubled and transformed existing spatial templates. However, this is not a geography of migrant struggles but, rather, a gaze on the troubled geography of the spaces stemming from migrants' presence in space and from their uneven temporality. Indeed, if we follow the moments and the spaces in which migrants become 'recordable' to our eyes and to discursive practices, we come to remap the European space, pushing the mapping gesture itself to the very limits of representation: something escapes the cartographic order, and the elusiveness or irregularity of migrants' presence defies the temporal narrative of geopolitical maps. That said, it is important to note that it is not a cartographic game that sustains the question 'which Europe?', or better 'which map of Europe, today?' Instead, the aim is to take on migrants' 'spatial insistences' (Sossi 2013b) and migrants' uneven geographies to see which map of Europe comes out. This entails undertaking a twofold analytical gesture. On the one hand, it means highlighting the untenability of the geopolitical map of Europe at present, namely, its constitutive inadequacy regarding the effective European spatiality, starting from the consideration that 'the old Westphalian map is misleading and no longer provides us with the necessary orientation' (Kratochwil 2011). This 'spatial incongruence' emerges in a quite glaring way if we bring attention to migrants' practices and, together, to the spatial impacts of EU migration policies. And at the same time a gaze on migrants' scattered and elusive geographies allows unpacking the state-citizen couplet which underpins Western political spatiality (Galli 2001). In both cases the limits of representation in accounting for the spatial transformations underway stem out: something inevitably falls out of the map, and it is precisely this outside—or better, its limits—that this chapter tries to engage with. The chapter develops along two lines. Firstly, it deals with the uneven current geopolitical map of Europe and shows that the European geopolitical map is itself misleading, taking into account the effects of migration policies and migrants' presence. Then, it investigates the temporal-spatial narrative of the cartographic-political gaze, analysing how migrants trouble the 'citizen epistemology' and the related order of representation.

A NON-CARTOGRAPHIC COUNTER-MAPPING APPROACH

What I call here a counter-mapping approach to the cartographic order and imagination refers to an analytical gesture which engages with the very limits of (political) representation at stake in the attempt to 'map' the spatial turbulence generated by migrants' unexpected presence, or by their being 'out of place'. Therefore, 'counter-mapping' designates an analytical posture which looks at the processes of re-bordering from the standpoint of migration movements (how migrants generate and transform spaces), drawing attention to the spatial reshaping they engender. At the same time, it studies the mechanisms of governmentality focusing on the impacts and the effects they have on spaces and on migrants' lives. In this way, both the mechanisms of border enforcement and the emergence or the transformation of transnational spaces and zones of detention can be read in the light of the 'spatial disarray' enacted by migrants: when migrants' incorrigible presence (De Genova 2010a) exceeds or unbalances the rhythms and number of selected mobility and of the expected clandestinity, the existing spatial mechanisms of filter, capture and management are forced to invent new spatial strategies. The reason for positing the question of spatial transformations via migrations addressing the cartographic spatial and political representation lies in the complex game of visibility and invisibility at play in the *migration strugglefield*. From such a standpoint, an attentive gaze should investigate which subjects and which practices remain under the thresholds of visibility, and how migrants' opacity suddenly changes when migrants need to be monitored or captured: the strategies of visibility and invisibility played both by mechanisms of control and by migrants shape an unstable regime of (in)visibility constantly destabilized by movements, events and new political assemblages. This is a regime, I contend, that shifts, rearticulates and transforms itself according to a specific temporality, namely, the tempos of the uneven and elusive migrants' geographies—whose condition of visibility is strongly dependent on their legal status—as well as the tempos of the frantic transformations of migration policies. The migration regime, which often becomes a struggle over (in)visibility between migrants and mechanisms of control, appears as a temporal map—that is, an uneven map changing in time—but changing according to the erratic tempos of migrants' movements and the accelerated temporality of the politics of mobility. Therefore, framing the issue of migrants' spatial troubling in terms of counter-mapping allows us to bring to the fore and engage in this contentious strugglefield, highlighting the political stakes behind Europe's spatial reshaping which concern time and (in)visibility. However, the counter move encapsulated in the notion of counter-mapping raises questions related to the issue of representation and narration. Indeed, if the conditions and the price of visibility are to be cautiously considered in the case of undocumented migrants, the very cartographic gesture cannot be taken for granted or proposed as a political tool to mobilize without circumspection. Since an in-depth analysis of the limits of the cartographic gaze goes beyond the scope of this chapter, I limit myself to mentioning two main problematic issues that cannot be eschewed.

First of all, the question of representation that is at the core of any mapping gesture: migrants' elusive geographies in part escape the script of representation insofar as their presence is at times overshadowed by governmental politics or because they try to pass unnoticed or, finally, because their presence in space is often fleeting and unpredictable. Something escapes or tries to escape the regime of representation: this is the conundrum of any migration (counter-)maps. Moreover, the move of fixing subjects and events in space that pertains to the 'cartographic grasp' does not account for uneven migration mobility. For this reason, the 'counter' in counter-mapping does not intend to redouble the cartographic gaze, inverting its rationale. Rather, the term 'counter' refers to a critical engagement with the elusive migration maps: it is an analytical posture which takes into account and respects unrepresentable migrants' geographies and their spatial presence, resisting the cartographic gesture and the mapping gaze of tracing boundaries and making things visible. More than tracing 'another map', a counter-mapping perspective tries to scrutinize and invent non-cartographic practices that point to the spaces in which the governmental map of migration as well as the geopolitical map of Europe appear as untenable illustrative devices. Thus, it is not even in terms of a coherent counter-narrative that a counter-mapping perspective acts: there is no unitary tale to be narrated, since migrants' spatial turmoil and their erratic temporality undermine precisely the consistency of the linear and ever-present spatial narrative displayed in the geopolitical map of Europe. In this way, the notion of 'counter' in counter-mapping has ultimately two meanings.

On the one hand, it refers to an analytical gaze that charts the effects of migration governmentality and the spatial disruptions generated by migrants. On the other hand, it challenges the very possibility of mapping those spatial upheavals, pushing the representative devices to their limits. Instead, counter-mapping tries to unearth the places and moments of spatial disruption and spatial reshaping; and, simultaneously, it tracks down the ways in which the exclusive access to spaces is challenged by migrants and then reconfigured by migration policies. In this way, counter-mapping suggests that spatial transformations and upheavals cannot be fully grasped except in relation to migration: indeed, as many scholars have argued, the very existence and functioning of borders is co-implicated with the production and the government of migration—namely, practices of movement which are translated and framed as migration (Hess, Karakayali and Tsianos 2009; Mezzadra and Neilson 2013). Thus, migration becomes a fundamental lens to see against the light processes of spatial reshaping and spatial disruption. Conceived in this way, counter-mapping is framed quite differently than in critical geography's original meaning: as a matter of fact, counter-mapping has usually been presented as a political strategy for 'appropriating the state's techniques and manners of representation in order to re-territorialize the area being mapped', assuming that 'these mapping projects are subversive since they exploit the authority of cartography' (Peluso 1995) putting on the map phenomena and subjects that tend to be left unmapped or silenced.

However, it is important to stress that migration counter-mapping exceeds the supposed specific boundaries of migration policies: such an analysis touches upon a much broader contested frame than migration itself, or rather, than a gaze on migration enables us to highlight. Indeed, a critical investigation on migration and spaces cannot overlook the (post)colonial question in which the cartographic gesture has a fundamental role: as many scholars have shown, mapping has played as an epistemic device, a technique for building the 'geo-body' of nations (Whinchakul 1997) and for crafting it as an 'imagined community' (Anderson 1991). Along with that, the maps of the colonial period served to present the American and African spaces as virgin areas, blank maps to fill and spaces where cartography effectively played as a technique of appropriation and dispossession (Harley 1989, 2001). But it is not only a national order of the space that is performed and represented through maps: more broadly, it could be argued that the cartographic gesture empowers a governmental grasp on spaces, fixing and allocating subjects and movements through a certain spatial b/ordering. Instead, the challenge of taking migration as a vantage point relies also on the attempt to undermine the 'sovereignty pitfall' of mapping. This does not depend on a particular obsession with maps but on the consideration that the cartographic gaze historically dominates and permeates the Western governmental rationale (Pickles 2004). Moreover, the struggle over (in)visibility that maps open up, selecting what can be represented on a map and what remains under the thresholds of visibility—the 'silences' of maps—inevitably brings migration to the foreground: indeed, migrants' strategies are substantially played out around the issue of visibility and invisibility, by struggling, escaping or negotiating the conditions of being *on the map*. The complex regime of (in)visibility that maps display and the geo-body of nation that cartographies not only represent but also, by presenting them as already there, contribute to establishing, leads us to the issue of representation as the *trouble spot* of any mapping gesture. This addresses simultaneously two aspects, within and beyond the cartographic order. It is quite intuitive to say that representation is constitutive of mapping: although maps do not simply represent the territory but have a performative and legitimizing function regarding border tracing, working as a kind of 'juridical territory' for conquering and governing spaces, and, as Korzibsky reminds us, 'the map is not the territory' (Korzibsky 1933), it remains that the political function of maps is always conveyed by the purpose of representing (something in) space. Such a discourse can be translated in the following terms: there ought to be no political science without border studies, and likewise, no border studies without mapping. But what is of interest in the perspective of a counter-mapping approach is the way that mapping and going beyond mapping converge around the issue of representation, or better, where mapping exceeds the cartographic dimension.

The theoretical and political relevance of maps relies precisely in the way in which they bring to the fore, foster and make visible stakes and premises that are at the very core of Western political thought and narrative. Indeed, representation is the principle which underpins, in an open or more implicit manner, the conceptual space of political thought, also tracing the boundaries and thresholds of what must remain *off the map*. A dissident gaze on maps—meaning by that a gaze which in part alters and resists the epistemic codes of mapping, forcing them into an alternative regime of knowledge—makes us pay attention to the implications of representation as a political and epistemic paradigm: the mechanisms of 'othering' that, as Henk van Houtum explains, are constitutively related to the process of bordering and are also peculiar to the working of political categories like citizenship (Van Houtum 2002). In fact, the production of subjects is traditionally conceived in conjunction with and by the condition of mechanisms of representation. Subjects are re-presented in public space, or become citizens and subjects of right: political imagination anchors the emergence of the subject to its 'authorized' and legitimate presence in space—that is to say, to a juridical supplement which makes subjectivity always located in a specific regime of representation. Along with this, as Étienne Balibar shows, in modern Western thought the citizen and the subject are postulated as inseparable political categories; and universality among subjects is ultimately based on the presupposition of an egalitarian sovereignty of citizens (Balibar 2011). However, far from being universal, the condition and the possibility of citizenship are grounded not only on the tracing of different degrees of non-citizenship but also on the fact that something remains under thresholds of representation. In this regard, a counter-mapping approach which works at the limits of representation by following migrants' uneven geographies forces us to rethink what could effectively mean a politics out of and beyond representation. In other words, the focus on the 'impossible cartographies' to trace migrants' elusive spatial upheavals is precisely what makes us connect cartographic and non-cartographic regimes of politics and mapping, positing representation as the 'trouble spot' to take on for rethinking political practices. Looking at maps beyond the map-text to see how migrants crack the order of the political, this methodological move ultimately encapsulates the meaning of counter-mapping that is situated just at the crossroads between maps and political thought, or better, at the junction where the cartographic and non-cartographic overlap, exceeding and blurring their own boundaries. To put it differently, the continuum between the cartographic order and the space of the political resembles what Graham Huggan defines as the correspondence between technical cartography and imaginative cartography (Huggan 1994). It is precisely by engaging at the limits of cartography that the political implications of maps come out and that, in turn, the rootedness of political hallmarks like representation become visible and translated in different linguistic and visual codes. Migrants' uneven geographies and the upheavals they produce allow shifting from the visual and political order of representation to the spatial transformations they engender, and to an enacted politics of presence that cannot be charted or narrated on maps.

EUROPE 'AT A DISTANCE' AND PATCHY EUROPE

The spatial disruptions and the reshaping of Europe that has characterized the last two decades have been hugely scrutinized by many authors, especially in terms of spatial rescaling (Brenner 2004; Mezzadra and Neilson 2013; Sassen 2013). And all these analyses underline that subjects, policies and mechanisms related to mobility have played a crucial role in the spatial re-bordering of Europe: the politics of externalization, the increased presence of migrants, the activation of the visa system and of bilateral agreements with third countries have caused such a substantial dislocation and fragmentation of Europe that it is not even possible to establish where Europe is. The image of a Europe 'at a distance' (Casas-Cortes, Cobarrubias and Pickles 2013) tries to capture the effects of the politics of externalization and of the Neighbourhood Policies that proliferated in the last decade. However, Europe *at a distance* can be a misleading image if confused with a territorial dislocation or enlargement of Europe: actually, the purpose and the effect of Neighbourhood Policies is not to integrate the third countries involved within the European Union but rather to make some administrative measures (like border patrolling against 'illegal' migration) and economic standards work there. Thus, Europe *at a distance* concerns less the territorial transformation than the reshuffling of Europe's spatial effects and influences. In fact, with regard to spatial transformations, it enables us to decouple space and territory, and also to disconnect sovereignty from territory. For instance, the combined Italian-Libyan patrolling along the Libyan coasts, established by bilateral agreement between the two countries, is less part of a direct territorial control strategy than of spatial reconfiguration of Italy's border management. Or in the case of the agreement signed in June 2013 between Morocco and the EU, the contested clause concerning the repatriation of third-country nationals on Moroccan territory—undocumented migrants deported from Europe—cannot be analysed in terms of territorial influence. Instead, the opening of detention centres for deportees, which worries many human rights organizations, will be one of the most tangible spatial transformations in Morocco engendered through the agreement with the European Union. Therefore, the spatial transformations in question and the spatial fragmentation of Europe are not necessarily something happening within the geopolitical boundaries of Europe: rather, the European space is formed also by non-territorial processes of bordering through which practices of mobility are partitioned, managed, monitored or detained.

Moreover, the image of a Europe at a distance requires a further consideration of the spaces that emerge from the discrepancies between the European Union and some member states. Indeed, in those cases it becomes impossible to trace a unique map: the emergence of a contested zone of action entails that different and in part conflicting maps of that new space of government can be narrated. One of the most incisive cases concerns a space that is not fully part of the European territory—the Mediterranean Sea—and the double governmental function deployed—rescuing and monitoring migrants' vessels. As I will show in detail in chapter 5, the Mediterranean Sea, despite the fact that it is not within the European territory—rather it is in the zone of international waters—has become a space of governmental intervention where competing legal frames and political responsibilities overlap. In the face of Europe's 'spatial conundrum' that makes it problematic to determine where Europe starts and ends (Walker 2000) and more broadly, where Europe is, the geopolitical map falls apart.

The recent bilateral agreements[1] between Italy and Libya concerning the deployment of an electronic frontier[2] along the Libyan coast to block the departure of migrants' boats is quite indicative of the way in which the enforcement and the proliferation of borders also cannot be fully analysed through the frame of territoriality. In fact, the goal of this new mechanism of bordering is less for Italy to exercise control over the Libyan territory than to force Libya to manage and block migrants' routes. In other words, it is a form of control which aims less to govern a space than to manage movements which trouble the order of mobility. Moreover, the agreement also entails the training of Libyan coast guards by Italian authorities, both in Libya and in Italy, like a sort of 'pedagogical sovereignty' practice made by another state (Italy) within its own territory and abroad.[3] In other words, these are scattered exchanges of sovereignty, beyond territorial borders—Libyans trained in Italy, Italian boats used for patrolling Libyan coasts, Libyan military personnel serving on Italian boats. It is crucial to remember that the colonial legacy still acts as a fundamental yardstick in the bilateral relations between Italy and Libya;[4] for this reason the exchanges in question cannot be analysed simply at the level of state relations but need to be complicated with the consequences of the colonial power relations. However, the Italian bilateral cooperation in part overlaps with the European program for Libya, called EUBAM Libya—approved by the European Council on May 22, 2013—that establishes collaboration by European military forces in patrolling the Libyan borders including some military units deployed by Italy. The cost of the operation is about 30 million euros per year and its main task is to advise, train and monitor Libyan authorities to build border management strategies in Libya and in Libyan national waters.[5] Therefore, both the variety of controls (shared information, joint patrolling, training of military forces, satellite surveillance) and the overlapping of European and national cooperation projects with complementary and in part conflicting competences hamper the envisioning of a stable map of the 'scattered Europe' produced by Neighbourhood Policies and bilateral agreements; or at least, no map of the scattered Europe seems possible on the basis of the spatial coordinates embedded in geopolitical maps. Overlapping joint-sovereignties (joint border patrolling) and de-sovereignty mechanisms (namely, EU and member states that 'train' foreign national authorities in order to teach them the best practices of control) constitute a patchy map formed of multilevel spatial scales of intervention.

In this regard, Foucault's definition of government helps us to better define the kind of action envisaged by this policing at a distance (by Italy, in this case) that is, however, based on joint-operations between European states and third countries: the idea that government means 'an action upon an action, on existing actions or on those which may arise in the present or the future' (Foucault 1982, 789) aptly describes the spaces on the move that this politics of control tries to manage. Indeed, the space of intervention turns out to be constituted by the very movements and routes enacted by migrants, trying to block or push them back. In fact, the expression *space on the move* makes us shift the attention from spaces to their transformations as contested sites of mobility and, ultimately, for their becoming borders. In particular, what needs to be considered is, on the one hand, the increasing bordering of the Mediterranean Sea concerning also international waters—namely, the multiplication of policies and treaties for governing movements at sea, rescue operations and mechanisms of control—and on the other hand, the normative voids and the legal quarrels between states on the right and the duty to intervene. All this depicts a space of government in which territorial sovereignties, policing at a distance (through radar and satellite systems) and the production of new spaces of governance like the limits of national waters and beyond, articulate each other.

THE GEOPOLITICAL MAP FALLS APART: MIGRANTS' UNEVEN GEOGRAPHIES AND THE PATCHY EUROPE OF MIGRATION POLICIES

From a counter-mapping standpoint, the spatial reshaping that migration policies and the politics of control have generated should be read in the light and against the backdrop of migrants' spatial disruptions, as well as of their uneven geographies that, as I will explain later, upset the temporality of the mapping narrative. To put it differently, a *patchy Europe* emerges from the spatial effects of the array of policies for governing mobility (politics of externalization, Neighbourhood Policies, new detention zones and mechanisms of remote control) in counterbalance to the erratic presence of migrants. In order to highlight some of these spatial turbulences, I draw the attention to migrants' movements and migrants' presence in Europe in the aftermath of the Arab revolutions. Milan, Paris and Marseille, 2011–2012: Tunisian migrants, who arrived in Lampedusa after the outbreak of the Tunisian revolution, troubled the order of mobility in which both authorized and unauthorized migrations are supposed to be situated. Indeed, they neither claimed asylum nor wished to stay in Italy; rather, they wanted to move away, to go to northern Europe and especially to France, without demanding international protection. For our purpose it is important to notice that they did not claim rights: their unexpected presence was characterized not by claims they made but by the way in which they persisted and moved in space, irrespective of the time and the conditions established by the politics of quota and the politics of selected mobility. They moved and stayed: some of them were blocked at the border with France and occupied public buildings in the cities of Padova, Bologna, Milan and Rome; others took their space in train stations or in public gardens. Some of those who arrived in France gathered themselves in 'collectives', collectives of 'shared geographies': *Le Collectif des Tunisiens de Lampedusa à Paris*—which was set up in Paris between May and August 2011 in the neighbourhood of Belleville—was formed by Tunisians who had arrived in Europe just after the revolution and who remapped their migratory experience stressing the common place of transit (Lampedusa) and the present 'unauthorized' location (Paris) (Sossi 2012). After the last eviction in August 2011, no trace has been left of that collective: on the trail of them both in November 2011 and in summer 2012, it was not possible to find any of the persons involved in the collective, although in the Tunisian squats in Paris a few remembered the collective, telling us that many Tunisians had been deported or had gone back because of the economic crisis in Europe.

But beyond the experience of self-organized collectives, the uneven geographies enacted by most of the Tunisian migrants did not appear in the form of a visible collective 'on the map' nor of an identifiable political subject. Rather, they played *off the map*, namely, under the thresholds of any traceability and political recognition; or better, their presence was visible at time intervals, depending on their juridical status, on their migratory projects and on the alternated migration regime of capture—sometimes more slackened, sometimes tougher, both in relation to the possibility of staying in a visible place although undocumented, as well as to the identification procedures, such as fingerprinting. Between 2011 and 2012, not only Tunisians but also 'Libyans' arrived on the Italian coasts: 'non-Libyan' Libyans, that is to say, migrant labourers who had worked and lived in Libya for years and who, due to the outbreak of the Libyan war, escaped to Italy. The North Africa Emergency declared by the Italian government in February 2011 with the arrival of the first 'Libyans' has certainly produced a deep spatial reshaping of the Italian system of detention and hosting: new facilities were opened and people were dispersed over the territory according to the mechanism of 'spread hosting', as defined by Italy—which also involved small facilities in all the Italian regions and located very far from urban centres. A variegated geography of centres and structures was activated, giving rise to highly non-homogeneous procedures and conditions of hosting. In fact, the North African Emergency was one of the most visible and politically immediate examples of bank effects and institutional reshaping generated on the northern shore of the Mediterranean. However, the mechanism of 'spread hosting' does not fully describe the effective spatial re-bordering which was at stake. Indeed, many protests and struggles occurred against the slowness in processing their claims of asylum and against the system of 'spread hosting'—which relegated some migrants to facilities in the Alps or left them without legal assistance. Beyond that, the mechanism of 'hosting' was in part bypassed by some of the 'Libyans', who tried to self-organize in the Italian town, squatting in buildings, or to look for jobs as hired hands in the country. From a counter-mapping perspective, it should be noticed that the official end of the North Africa Emergency—31 March 2013—did not correspond to the disappearance of migrants' presence on the territory. On the contrary, their spatial visible presence has become also more discontinuous and unpredictable. The measures adopted by the Italian government for chasing away 'Libyan'[6] migrants—500 euros per person to leave the facilities—were not accepted by all of them; due to the economic crisis in Italy, many of those detained in the biggest Italian hosting centre for asylum seekers (Mineo, Catania) decided to stay in the centre to at least get food and shelter for free. Others self-organized in the Italian towns without following the established patterns of integration for refugees, which many of them considered largely inconsistent. In the place of the humanitarian logic of hosting, 'Libyan' and Tunisian migrants tried to take and make a space for themselves, out of the exclusionary humanitarian channels of the asylum system.

AN INTERMEZZO ON MIGRANT STRUGGLES AND FREEDOM OF MOVEMENT

Migrants' refusal to be governed in that way shows indeed that migrations are always crisscrossed by and enmeshed within struggles; that is to say, it is not just when migrants engage in an organized struggle or lay claims that the conflicting dimensions of migrations come out. This doesn't involve a romanticism which sees migrants as struggling subjects; on the contrary, the idea that particular struggles and resistances against specific instances of non-functioning of the asylum system actually convey a broader disqualification and a refusal of the ways in which migrants' lives are governed enables us to downplay the stress on visible and organized struggles. In fact, being called a migrant involves being captured and shaped by a complex set of policies, knowledges, techniques and encounters that make migrants' lives governable *as migrants*. The *struggles for movement* that unauthorized migrants engage in beyond public demonstrations and visible violations of borders are strategies of mobility and strategies of 'spatial insistence' through which migrants can successfully avoid being blocked or pushed back. Strategies of flight and strategies for persisting in a certain place are struggles over space that migrants ordinarily engage in and that fall out of the traditional struggle-form that activists usually replicate. The formula 'migrant struggles' encapsulates at least two different meanings and refers to an array of different empirical experiences of migration. First, 'migrant struggles' is the name for the multiple concrete struggles in which migrants are engaged: more or less organized struggles that defeat, escape or trouble the order of mobility; struggles taking place at the border or before and beyond the borderline; struggles that gain the scene of the public space or that remain invisible. In this first empirical meaning, the notion of 'migrant struggles' unpacks and pluralizes the catchword of migration, highlighting the heterogeneity of migrant conditions and the different ways in which migrants are confronted with powers. However, taking migrant struggles in this first empirical meaning, we should resist seeing any practice of migration as deliberate agency or as a challenge to the border regime. Rather, it is important to keep in mind the ambivalences that cross the practices of migration: they could play as resistances against some mechanisms of containment or against the social norms of the country of origin while at the same time being the other side of the mechanisms of selected mobility that channel authorized movements. From this point of view, the concern becomes one of bringing out what exceeds the economic frame in which migrants' presence is required and expected. But along with this first empirical meaning, 'migrant struggles' indicates that every migration is situated in and grapples with a certain strugglefield, and in this sense is always crisscrossed by and involved in multiple struggles, as captured, filtered and managed by techniques of bordering and control. It follows that migrations are eminently implicated in relations of power and conflicting fields of forces; consequently, any migration as practice taking place within such a strugglefield is immediately also a struggle for modifying, challenging or interrupting that configuration of power. And, at the same time, migrations force the border regime to constantly revise its strategies—in a way, working as a constitutive troubling factor. After all, by naming these migrants' struggles as *discordant practices of freedom* what is highlighted is also the specific freedom that migrants enact when they move or stay in space despite the techniques of bordering that set the pace of the terms and conditions of mobility.

More than focusing on a geographic area—Italy—fixing in advance the boundaries of the impacts of that spatial reshaping, I switch the attention to the spaces that emerged or were transformed across national borders due to migrants' uneven presence and to the restructuring of migration policies. In other words, a possible alternative spatial gaze consists in accounting for these processes of spatial disruption by following migrants' uneven geographies in conjunction with the transnational impact of migration policies. Taking together the spatial effects of migration policies and migrants' spatial turbulences, what emerges is a space that only in part overlaps with the geopolitical map of Europe and its boundaries. The rescaling and reshaping of borders across Europe concerns only in part visible borders: if, on the one hand, detention zones and refugee camps can be considered bounded spaces marked by acts of border tracing, the emergence and the effects of other spatialities remain undetectable to the cartographic eye. For instance, the spaces of mobility produced by bilateral agreements signed with third countries are difficult to chart, since those spaces do not necessarily correspond to new territorial divides but rather are actualized through the routes and the movements of selected mobilities established and allowed by those agreements. Moreover, these spaces have a specific and uneven temporality which cannot be compared with the linear tempos at stake in the geopolitical maps: migration policies instantiate *conditional spatialities*, that is to say, spaces that exist only for some categories of mobile people or that are accessible to some only at intervals. The most pertinent example is given by the Mobility Partnerships that are based on channels of selected mobility to which some people can—sometimes only temporarily—get access.

Indeed, as Mezzadra and Neilson point out, the supposed sharp distinction between skilled and unskilled migrants is actually embodied by many migrants, whose labourer status fluctuates in time (Mezzadra and Neilson 2013). It follows that the migrant condition needs to be pluralized and diffracted, not only case by case but also concerning a single subject: the condition of being a migrant as well as the manifold migration categories are experienced and enacted in different ways by people from time to time. Therefore, without overlooking the proliferation of internal walls across Europe (new border zones and detention areas), from a counter-mapping standpoint the analytical gaze should be directed towards borders which remain invisible on the maps: borders which generate or foster differences in space—for instance, by putting into place a differential access to space—and borders which ultimately coincide with managing movements governing the routes and the speed of mobility.

MIGRANTS' ERRATIC PRESENCE AND THE TROUBLING OF THE CARTOGRAPHIC SPACE-TIME: LAMPEDUSA, HAMBURG AND CALAIS

The spaces emerging from migrants' movements or from migrants' unexpected stays can hardly be accommodated or represented on a map. If one follows the erratic geographies of Tunisian migrants across Europe between 2011 and 2012—whose fragmented journeys were fundamentally made of round-trips between Italian and French cities—their elusive character makes them impossible to trace as lines on a map. And if one also takes into account migrants' visible claims and struggles, all these (in part) escape the cartographic order: the spaces taken or named by some migrant collectives, like the *Lampedusa in Hamburg*, do not correspond to any zone or line that can be traced or located on a map, nor can they be envisaged in terms of 'borders'. Rather, those migrants' struggles make certain spaces—that can also exceed localizable areas—into spaces of stay, re-signifying or reshaping the space in the light of migrants' enacted geographies. The *Lampedusa in Hamburg* refugees' group neither designates nor simply renames the city of Hamburg as such, but talks about 'a space in action', namely, the space enacted and travelled by those migrants who arrived in Lampedusa in 2011 and then, after obtaining the temporary permit decided to move to Germany because of the economic crisis in southern Europe.[7]

Similarly, the 'Syrians blocked in Calais', as the Syrian asylum seekers who wanted to go to the UK defined themselves, refer to the city of Calais in terms of the ways in which it limited, fragmented, impacted on and related to their journey. Syrian migrants neither arrived in Calais all together—although all of them arrived in the first half of October 2013—nor did they come by the same route—some arrived via Turkey, others from Libya and Lampedusa. They became the group of Syrians stopped in Calais soon after the blocking of the ferries to Dover on 2 October for two days: they went up to the roof of the harbour departure zone as a protest against the treatment they received in France and demanded that the UK accept their asylum claims. The protest had a huge resonance in the media, since it was one of the few protests in many years in Calais. And when British authorities came to negotiate with them, this was seen as an unexpected victory. However, as usual, after a few days no news or update circulated any more on the web about the outcome of the protest. When media and political attention to migrant struggles stops, migrants' spaces become nonexistent in the eyes of the order of political visibility. Indeed, the temporality earmarked to migrants' spaces is ultimately the temporality of the narration on them, which coincides with the irruptive moment of the political 'emergency' or of migrants' visible struggles—namely, the moment when migrants come on the stage and there are some 'events' to be narrated. I arrived in Calais fifteen days after the end of the struggle, just in order to understand what was happening to those Syrians and the outcome of the protest. The sixty Syrians who staged the protest were still in Calais, with their tents near the harbour, but far from considering the struggle a victory, they stressed that British authorities instead succeeded in neutralizing their claims: 'Actually, British politicians came here just to explain to us the rules of the game, namely, the UK's mechanisms of asylum and did not open any concrete possibility to us. Thus, they voided all our claims, since at that point any further claim appeared useless'.[8] However, despite the existential and juridical impasse, they decided to write a collective letter addressed to the European countries, demanding the possibility of settling and claiming asylum in any country without being forced to give their fingerprints in the first European country in which they arrived—a measure which went against the Dublin II regulation. Coming back to Calais twenty days later, the space occupied by Syrians and their condition was further changed: some of them had managed to arrive in the UK, while the others were still waiting there, in the occupied square, and the composition of that space was transformed by the arrivals of migrants from Afghanistan and Iraq.

16 July 2013, Lampedusa: two hundred Eritrean migrants marched in the streets of Lampedusa, protesting against the requirement to give their fingerprints in Italy, claiming their willingness to move on and claiming asylum in a different country. The protest then became a sit-in in front of the main church in Lampedusa, and Italian authorities agreed not to take their fingerprints. During the demonstrations, with the main slogan 'no fingerprints by force', they asserted that, moving to northern European countries as they wanted to do, their physical presence should be disjointed from their digital data. However, after they were moved to Sicily it was not possible to know what had happened to them—whether they were fingerprinted or not once on the mainland in the detention centres or if Italy let them escape to Germany and France. Was it really a victory for them and had Italy again disobeyed the Eurodac regulation or, de-

spite the very important symbolic result (not being fingerprinted in Lampedusa) were they then identified in Sicily? To formulate at least a partial answer it was necessary to follow at a distance the traces of a few of them, who, once they arrived in Germany, came into contact with networks of activists. As two Eritrean migrants explained, although Italian authorities officially tried to identify them in the hosting centre for asylum seekers, actually they were not particularly concerned about them, de facto allowing them to escape.

In order to complicate the counter-mapping gaze on migrants' spatial upheavals, it is important also to come to grips with the temporal pace of migrants' geographies which conflicts with the temporal narrative of geopolitical maps. In this regard a look at practices of migration taking place in the aftermath of the Arab uprisings provides us with a useful lens for unpacking the temporal issue that is at stake in any map. In fact, three years after the outbreak of the Arab uprisings most of the 'Libyan' and Tunisian migrants are scattered across the European space or have come back because of the economic crisis in Europe. The political collectives of Tunisian and 'Libyan' migrants that emerged in 2011 and 2012 in some European towns do not exist anymore. Therefore, in the face of the elusiveness of the *spaces in action* re-signified by migrants, it is fundamental to investigate to what extent migrants' spatial disruptions succeed in initiating and establishing a political language which goes beyond the register of recognition without vanishing after a specific political struggle comes to an end. Actually, this appears as one of the most outstanding impasses, once considered the elusive character of migrant struggles and migrant spaces; and for this reason the effort of rethinking the political space beyond both the paradigm of representation and the traditional spatial representations seems to be the fundamental issue at stake. Drawing on Carlo Galli's argument about the relationships between politics and space, any political space and thought is sustained by specific spatial representation, and in the case of modern Europe these 'concretely organize the spaces of freedom and citizenship' (Galli 2001). Therefore, focusing on migrants' uneven geographies means paying attention to the way in which established political categories and spatial coordinates which structure our geopolitical imagination fall into crisis.

However, instead of focusing on a given strugglefield—a space of struggles in which politics of bordering and migrants' strategies come into conflict—a dislocating move can be undertaken: rather than the duration of a movement or struggle, we can scrutinize the circulation and dissemination of a certain language that can be put at the core of the analysis. In fact, engaging in a counter-mapping approach means to challenge the supposed fixed spatial narrative which traditionally underpins political imagination and language (Van Houtum 2005). In other words, the issue of political subjectivity and language can be reframed, reformulating it according to a mobile spatiality which corresponds to the dissemination of some practices at a distance. This entails disengaging from the 'territorial trap' (Agnew 1994), in which many analyses of migrant struggles are caught, and stressing their territorially based character. Without denying the importance of the specific place in which migrants' geographies are enacted, I suggest bringing attention to the way in which those local and territorial struggles or movements contribute to the emergence of 'spaces in action' that do not simply cross the borders. Indeed, more than thinking in terms of crossing (between geographic or political distances), we might look at those spatial upheavals as the production of interconnecting spaces. And more than questioning their ability to last in time, we might turn to their reverberations at a temporal distance. The spaces that migrants transform cannot always be marked or fixed on the geopolitical map—towns, nation-states, border-regions—nor can they be narrowed to the act of border crossing. Rather, migrants' uneven geographies enact/transform spaces which discordantly overlap with the existing geopolitical map.

Even if focused on migration, the counter-mapping approach I mobilize enables us to highlight broader spatial practices that, despite being located in a specific space, are not territorially bounded and clash with the temporal order of maps based on a linear passing of time and on a lasting persistence in space. To phrase this in other terms, I suggest that counter-mapping engages in the effort of not fixing in advance the spatial units and spatial scales on which migrants act. In fact, analyses of migrant struggles tend to take for granted the spatial unit in which practices of migration take place, reiterating a sort of 'spatial mastering' on migrants' movements: instead of taking and rethinking together migration and borders/spaces, practices of migration are ultimately flattened and encoded through existing spatial analytics, hampering the possibility of seeing the upheavals that migrants sometimes generate. Therefore, not only do migration and border-spaces need to be analysed together, but how migrations impact and transform spaces and how new space-time units are put into place should also be investigated. From this perspective, instead of addressing migrants' presence and action in space through the bounded spatialities in which we usually move, we ask which spaces are produced or enacted by migrants' movements. It is precisely in this sense that migrants unearth territorial units: not because they do not act in a local space, but because they enact geographies that exceed and trouble any possible 'methodological nationalism' (De Genova 2010). Above I argued that the *Lampedusa in Hamburg* collective is not merely a group acting and living in the town of Hamburg. Or rather, it is that and yet at the same time it cannot be really framed through the given space of that town: actually, 'in Hamburg' designates the place of arrival and, in particular, the specific squares, the occupied buildings and the streets where refugees live. But the location of Hamburg is complicated by the journey narrative through which they name their own space: *Lam-*

pedusa in Hamburg evidently does not correspond to any place 'on the map', but it is an enacted space that those migrants travelled and in some way produced. Simultaneously, Hamburg refers to a claim, the claim for a space that is not already there: a space for them to stay—since the municipality of Hamburg refuses to recognize the rights they have been entitled to as refugees in Italy, as well as the right to stay. However, there are also more hidden spatialities that are produced and enacted—or produced *as* enacted—by migrants: the uneven geographies of the Tunisian migrants who scattered across Europe for months before settling in Paris or being deported or deciding to go back to Tunisia are spaces travelled that are not bounded by the spatial-units of the geopolitical map.

The *Lampedusa in Hamburg* collective, the *Syrians blocked in Calais* and the two hundred Eritrean migrants refusing to give their fingerprints in Lampedusa—these three snapshots correspond to episodes located in European places that, at the same time, were produced in *spaces on the move*, spaces of moving and staying that do not fully correspond to the geographic locations named here. These snapshots also show the uneven temporality of migrant struggles and migrants' geographies: a political collective based in Hamburg characterized by a common place of crossing (Lampedusa); a group of Syrians who for about one month remained blocked in Calais, staged a protest and then, after waiting for an extended time, scattered across Europe—some arriving in the UK, others applying for asylum in France and some going to Sweden; and a huge two-day protest in Lampedusa made by Eritreans who resisted giving their fingerprints, aware of the Dublin II regulation but who after their successful claim, were taken to the mainland where their traces were lost. Therefore, behind the label 'migrant struggles' a heterogeneity of tempos crumbles this supposedly stable referent, forcing us to a rethink politics and political practices, starting from the multiplicity and the unevenness of migrant conditions. In this way, these different snapshots show that there is not a coeval 'time of politics', and meanwhile they enable us to see that there is not even something like a 'migration tempo' to oppose to the supposed linear progressive time of citizenship. Rather, 'migration' as a political and analytical category diffracts into different but coexisting temporalities.

COUNTER-MAPPING AT THE LIMITS OF (THE POLITICS OF) REPRESENTATION

The fundamental non-synchronicity of the many 'times of politics' at play in the European space of migration and the uneven maps that emerge resound in the spatial and temporal cartography of colonial modernity (Goswami 2004). In order to understand the ambivalent relationship between diaspora time and the West and the tensions between different coexisting temporalities, Paul Gilroy suggests that a 'stereoscopic sensibility' is required 'adequate to building a dialogue with the West: within and without' (Gilroy 1993). Translated into the contemporary European space as a space of migration, this counter-mapping approach is situated at the crossroad between, on the one side, the production of an alternative cartography from within the existing geopolitical coordinates, and on the other, the limits of representation as such, namely, the gesture of resisting mapping migrants' upheavals. To point to the limits of any possible representation means also to take into account how migration policies and migrants' upheavals impact on existing spaces and, simultaneously, how they produce other spatialities that partially overlap and clash with the existing geopolitical map of Europe. To pay attention to the 'tension between roots and routes' in the map of the contemporary 'scattered Europe' entails making migrants' persisting in spaces resonate with the production of *discordant spatialities*, namely, spaces which do not match with the temporal pace of the politics of mobility. In other words, migrants' upheavals must be situated within existing spatial coordinates and scales and, jointly, how these latter are displaced by them must be examined.

By taking into account the spatial effects produced by the EU's politics of bordering in third countries, and at the same time the spaces upset by migrants' movements in the European territory, I tried to overturn the usual spatial fixes that are sustained by a cartographic gaze and that imprint our spatial imagination. An analysis of 'scattered Europe' engendered by the EU politics of mobility and the *spaces on the move* produced by migrants shows that the geopolitical map of Europe conceals multiple spatialities and spatial scales simultaneously at stake, staging a dominant spatial unit. The European space as pictured in cartographic maps is disturbed by the patchy *spaces on the move* that migrants enact. However, in this chapter I also stressed that a counter-mapping approach does not aim to produce contrapuntal maps, tracing migrant cartographies to oppose to the supremacy of geopolitical maps. Instead, 'counter' in counter-mapping means also situating subjects at the very limits of what can be represented, refusing to encode any subject and movement within the cartographic order. To conclude, a further consideration is needed on *Europe in migration* and on *Europe of migrations* which these snapshots have shown. By drawing attention to migrants' spatial upheavals and to the scattered figure of Europe, the issue is not (only) to take the migrants' perspective for looking at borders and spaces, engaging us in a sort of reversing gaze—how migrants see borders—but rather to keep up with the effective spatial transformations they generate. In other words, a counter-mapping approach follows the spatial effects and the turbulences that migration policies and migrants put to work, making geopolitical Europe collapse in patchy and clustering spatialities. The map, as an organizing principle of spatial and political imagination, starts to break down in the face of migrants' unmappable spaces. In the end, an insight into migrants' spatial upheavals makes it possible, as Harley puts it, 'to read between the lines of the map' (Harley 1989).

THE MIGRANT-SUBJECT BEYOND THE SNAPSHOT

However, taking snapshots of specific spaces and moments of migrant struggles confronts us with the uneven temporality of the migrant condition, and with what I call the struggle-form. In fact, as shown in this chapter, the 'irregular' condition of the migrants arrived from Libya entails an elusiveness of their presence in the public space as well as a transformation of the migrant condition itself. The moments in which migrants use strategies of visibility—making space for themselves in the cities, claiming rights or making protests—are actually alternated with periods of invisibility. In this regard, the analytical gesture of the snapshot technique should avoid reproducing spatial fixes that look at struggle-forms that pertain to the usual grammar of political mobilizations. Migrants also struggle and enact strategies of movement in ways that cannot be frozen into the space-frame of the momentary and visible struggle that addresses state institutions or governmental agencies. Related to this point, the migrant-subject that snapshots focus upon actually represents only a temporary subjectivity of a 'subject in transit' (Mezzadra and Neilson, 2013) and is not comprehensive of the full migrant condition that a person experiences across spaces. In fact, the 'migrant condition' that one experiences is actually formed by different and constantly changing juridical statuses, subjective interpellations and existential conditions. Therefore, beyond the snapshots that 'capture' migrants at a particular moment of their journey and of their migration status as well, it is important to draw attention to the impact that different kinds of borders (geopolitical borders, juridical borders, racial borders) have on people's lives. More broadly, the 'becoming migrant' phase cannot be hypostatized into a given migrant condition: rather, it is the outcome of juridical and administrative measures, migration policies and labour policies that shape different migrant conditions. Moreover, any migrant-subject is subjected to a series of identity reshufflings in response to changing juridical and political circumstances, as well as the substantial transformations related to being a migrant in different spaces.

NOTES

1. The most recent cooperation agreement was signed by Libya and Italy on October 28, 2013.
2. The electronic monitoring system to be implemented on the Libyan coast consists of a radar network, provided by the Italian company Finmeccanica for an amount of 300 million euros, and a system of air surveillance.

3. The Italian military mission in Libya established that Italy would provide the Libyan authorities with technical equipment, armaments and vehicles. At the beginning of 2013 Italy had officially allocated 7.5 million euros for this military cooperation, which included the maintenance of the Italian navy units of the Guardia di Finanza given to Libya.

4. The bilateral "Friendship agreement", as it has been called, signed between Italy and Libya in 2008 established that Italy would give Libya 5 billion U.S. dollars per year for the next twenty years as a form of colonial compensation.

5. EU Border Assistance Mission (EUBAM) in Libya, http://www.eeas.europa.eu/csdp/missions_operations/eubam-libya/eubam_factsheet_en.pdf

6. Actually none of them were Libyan citizens; the people who fled Libya by boat and arrived in Italy were all migrant workers—most of them settled in Libya for extended periods—but they were very often presented by Italian media as Libyans.

7. http://lampedusa-in-hamburg.tk/

8. Interview conducted on 10 November 2013 with two Syrian asylum seekers in Calais.

FOUR

Democracy as a Strategy of Containment and Migration in Crisis in Revolutionized Tunisia

The political upheavals which took place in 2011 in many Arab countries were immediately depicted as the awakening of the Arab region; the trademark of 'Spring' became a signifier of the belated race to democracy that Arab people undertook in struggling against dictatorship (Badiou 2012; Eyad 2012). The Arab Spring was situated into the historical prose of the Enlightenment—the delayed Enlightenment of the Arab countries—and the Mediterranean was presented as a space for a fruitful political dialogue between the two shores: 'as a result of the recent Arab mass uprisings, a new Mediterranean is emerging' (Ammor 2012, 128). At the same time, the 'Arab' label encompasses different contexts of struggle through an indistinct signifier. The rallying cry of freedom and democracy resounded across the Mediterranean, with both liberal analysts and left-wing movements looking at those revolutions as a promising prospect for political change and as liberation from dictatorships. Political analyses on the Arab uprisings from the northern shore of the Mediterranean corroborated what appeared as the 'unavoidable path to democracy': in fact, the democratic script representing the universal horizon of different and also opposed political views constructed democracy as the most desirable political form (Chatterjee 2012).

The 'yardstick of democracy' contributes towards creating the image of a smooth and all-connected space, which slightly changes according to the political perspective one mobilizes—the international political space (Jabri 2013), the transnational space of mobility (Blunt 2007); and the global community or the global citizenship (Rygel 2011). From this point of view, if, as Benjamin argues, the notion of human progress is inseparable from the idea of history as a homogeneous and empty time, in a similar way the blueprint of democracy refers to and is predicated upon the image of a homogeneous space—not so much an empty one as, rather, a networked space. In other words, the global space is the correlate spatial referent of democracy and, conversely, democracy is the adequate political model for a globalized civil society that wants to overcome the social inequalities produced by globalization in favour of new spaces of participatory citizenship. But democracy is not spatially fixed: on the contrary, it is supposed to travel across spaces; it necessitates spaces and times of transition. The narratives of the social uprisings tell us that a real participatory democracy is produced precisely through the connection and mutual contamination between spatially dislocated struggles. Therefore, democracy first of all requires a ceaseless process of spatial, epistemic and political translation (Butler 2009; Irrera and Ivekovic 2013): the appropriation of spaces during the Arab uprisings and the occupations of public spaces in the Occupy movement; the language of precarity in the social struggles in Europe, and the 'noisy' London riots; the real democracy in the Acampadas movement and the democracy-to-come in the Tunisian revolution. However, what kinds of movements across spaces are encouraged, fostered and legitimized through the script of democracy? And if we take democracy not only as a discourse but also as a technology for governing subjects and populations, how does it articulate with (the government of) mobility? In fact, a few years after the outbreak of the Tunisian revolution, there are many analyses concerning the discourse on democracy, its normalizing function and the confrontation with European struggles for radical democracy (Balibar and Brossat 2011). Instead, what remain substantially unexplored are the democracy-mobility nexus and the working of the democratic norm not only as a discourse but also as an effective political technology that binds people to specific rules of conduct and autonomy. Hence, democracy works as a *discipline of mobility*—the democratic frame and standards are posited as the necessary conditions for enacting an ordered free mobility—and as the yardstick through which disparate practices of refusal and uprising against the government over lives are re-codified into a common linguistic and political space, the space of connections. The representation of a shared space made of multiple connections (networks, similarities in struggles, common desires, political compositions, etc.) and the immediate translation of emerging political subjectivities into marks of a renewed desire for an effective democracy, paradoxically, make democracy the normalizing monolingual code. This history of the present tries precisely to challenge the uncontested democratic norm, interrogating the new political spaces and subjectivities that the *twofold spatial upheaval* in the Mediterranean has opened up, which are not fully translatable into the given frame of democracy. As this chapter illustrates, democracy operates in revolutionary spaces as political technology for taming turmoil and as a strategy of containment. The turmoil generated by the Arab uprisings stretching well beyond national boundaries and the current political developments that are unfolding, from Egypt to Syria and Tunisia, can hardly be understood through the 'smooth narrative' of democracy. Thus, by troubling the tale of democratic transition, the Arab uprisings and migrants' spatial upheavals has interrupted the temporality of the 'not-yet' that pervades colonial and postcolonial spaces: the temporal border represented by the supposed progressive route towards democracy that put (post)colonial spaces simultaneously at a spatial and temporal distance from 'full-fledged democracies' (Chakrabarty, 2000b) has been vehemently broken up and reshuffled during the Arab uprisings.

The wave of upheavals was framed from different angles as a new open strugglefield for democracy. Nevertheless, this pleasant smooth tale was destabilized by the departure of thousands and thousands of Tunisian migrants towards Europe, and the conquests of freedom and democracy became a more ambiguous concern in the face of troubling mobilities. While not yet put into place, the new democratic space appeared to be a very unstable region, subject to diverse possible crises: the migratory crisis, the crisis of just-born democracies, the debt crisis and the humanitarian crisis at the Libyan border. The unexpectedness of migrants' departures was translated in terms of a predictable failure of a still unstable economic and political context trying to achieve an accomplished democratic system of governance. Migrations were seen as the undesired side effect and as the index of the political crisis springing in new-born democracies. Very quickly, the tale of revolutionary democratic upheavals slipped into a narrative of discontent and turmoil fostered by social and economic inequality, laying the ground for presenting it as a phenomenon to be governed and finally tamed.[1] Migrants' movements were depicted as the index of the political turbulence triggered by Arab uprisings, requiring a structured response by the EU: migrations were the troubling outcome of undisciplined democratic uprisings that produced 'deviant' impure models of democracy and 'disturbing' processes (Chatterjee 1993, 3).

In this chapter I bring to attention the spatial upheavals produced in the Mediterranean at the time of the Arab uprisings, focusing on the new spaces they produced and the spatial economies which they disturbed. I situate the analysis at the intersection between practices of movement and spatial economies, interrogating how different regimes of government, truth and mobility have been created, transformed or re-signified. In the first section I focus on the idea of democracy as a strategy of containment, analysing how the politics of mobility and economic projects of development articulate in revolutionized space through the script of democracy. Then, I take into account the notion of crisis as a catchword used by migration agencies for coming to grips with the migration turmoil that took place in the Mediterranean. Finally, in the third section I turn to the project of a Maghreb transnational area of free mobility, challenging the European referent as a space of free circulation. Hinging on the fundamental openness that characterizes these political events, I talk about *revolutionized spaces* in the place of a post-revolutionary context in order to mark the ongoing political turmoil taking place on the southern shore of the Mediterranean, stretching the temporality of the revolution fixed by the European narrative of the Arab uprisings.

The nexus mobility-democracy was mobilized simultaneously in two opposite directions. On the one hand, it was built on the concept of 'good governance': democracy is posited as a guarantee and a necessary condition for managing ordered mobility, while (selected) mobility is seen as a propelling driver for Tunisia to achieve a proper transition to democracy.[2] On the other hand, (disordered) mobility and (undisciplined popular) democracy were designated precisely as the causes of political instability in the Mediterranean. Since the 1990s, the migration-development[3] nexus has been presented as the cornerstone of Migration Studies and policies, conceiving migrations as a factor constitutive of developmental policies (Castells and Delgado 2008; Faist, Fauser and Kivisto 2010; Ghosh 2000; Hess 2008; Pastore 2007; Piper 2009; Tapinos 1990): migrations are managed as a developmental solution, and the main challenge of intergovernmental agencies becomes to fix subjects in space, preventing people from migrating in the name of developmental goals. At the same time, migration has started to become the issue through which developmental strategies and discourses are redefined. However, the migration and development blueprint ultimately relies on the ambivalence of the question 'development for whom?' In fact, two simultaneous political orientations are encapsulated in the formula of 'migration and development': on the one hand, a selected and managed mobility towards Europe for countering the European demographic crisis is seen as a 'developmental agent' (Faist 2008); on the other hand, developmental projects in migrants' countries of origin for tackling the 'root causes of illegal immigration' (Chaloff 2007). Therefore, if the former is commonly recognized as a 'development through migration' strategy, I would call the latter the rationale for 'non-migration through development'. The script of democracy is posited, on the one hand, as the pre-condition for promoting local development and ordered mobility; on the other, it is presented as a destabilizing factor boosting the social turmoil that needs to be managed implementing regulative measures and projects of development. Indeed, the Arab uprisings are staged by European analysts as a source of economic and political instability and, simultaneously, as an opportunity to put into place new economic agreements.

In revolutionized Tunisia and in Libya the migration-development agenda found a fertile ground, and at the same time it was destabilized by practices of migration that could not be fully regulated through the logic of 'development instead of migration'. The script of the transition to democracy was introduced through the development grid as a way of taming the turbulence of peoples' mobility and social unrest according to the logic of a temporal lag: the gradual conquest of democracy, the quest for secular values and the education in democratic best practice, are the three main pillars of Europe's discourse on the belated democratic arrival of people in the Arab world.

From the northern shore of the Mediterranean, the Arab uprisings were depicted as a considerable step forward for decreasing the distance and the asymmetry between the two shores: the Mediterranean emerges as a homogenizing spatial signifier, establishing a series of multiple proximities between Europe and the southern shore. Consequently, in the name of proximity, new exclusionary borders are traced out, excluding some countries and geographic areas from the logic of partnership as well as from the discourse of a common space to share and develop. Moreover, these borders are essentially blurred and always changing, since the Mediterranean space itself, as an economic and political landmark, is ultimately the provisional result of the combination of different agreements and networks—the Union for the Mediterranean, Mediterranean Dialogue 5+5, Euro-Mediterranean Human Rights Networks. Nevertheless, as soon as the attention shifts to migration, the distance between the two shores seems to increase: migrations are seen not as practices of freedom that are part of an effective democracy but rather as side effects of ungoverned social unrest and deceptive practices which try to evade the law.

THE PRODUCTION OF TEACHABLE SUBJECTS

In the aftermath of migrations' upheavals and revolutionary uprisings in the Mediterranean, migration governmentality has been rearranged, taking the migration-development nexus as its main linchpin. This is not at all an original agenda in migration policies—making people develop their country in order not to migrate—but what is peculiar is the way in which the strategy of 'making the poor work' and politics of migration are articulated to govern political instability, economic discontent and practices of mobility, surreptitiously positing mobility as a symptom of unrests and troubles. Political upheavals in Tunisia were sized as the opportunity to revise bilateral agreements and Mobility Partnerships; and conversely the migratory issue has become a constitutive part of the process of transition to democracy. Thus, more than setting autonomous developmental policies, international migration agencies place their interventions within wider economic projects of a 'struggle against poverty'. In this way, an in-depth analysis of migratory policies necessarily needs to take into account the economy of people's mobility at large, including economic projects of development in revolutionized spaces and the relationship between the labour force and the territory.

Developmental policies put into place strategies for fixing people in space, positing the migrant as a subject who must learn to fight poverty while remaining in her/his own country. In this sense, the economy of subjectivation in the postcolonial revolutionized space cannot be fully encapsulated into the logic of human capital, since many political technologies of capitalization of the labour force are simultaneously at play. After all, the model of human capital frames capital as a social relation within the kernel of the 'human', failing to account for the overlapping, and sometimes conflicting processes of subjectivation as well as of exploitation and de facto making the labour force as a social and economic category disappear (Mezzadra and Neilson 2013). I refer, for instance, to the dynamic of qualification and disqualification of people's skills based on the juridical and social position of subjects. Broadly speaking, the model of human capital is not fully adequate for casting light on the different mechanisms at stake in the moral geography of migration governance (Hyndman 2000). While *Homo oeconomicus* is a subject that responds only to its own interest and for this reason can be governed according to economic principles (Foucault 2010), would-be migrants in revolutionized Tunisia are addressed as teachable subjects who need to learn practices of democracy and to engage in an entrepreneurial rationality while remaining in their own economic and geographic context. What is at play is a multifaceted rationale in which the entrepreneurial principle, the learning of 'best practices' of democracy and the injunction to remain in one's own place coexist.

However, it is important not to follow closely the governmental narrative, remaining at the level of its texts. In fact, if we consider the developmental projects and economic activities promoted by international and European agencies in Tunisia, we see that they are far from fostering individual autonomy. What is effectively sustained and funded is a set of unskilled economic activities going in hand in hand with the idea of the inevitable failure of would-be migrants in performing a model of autonomous and self-improving subjectivity. Indeed, what Sanjay Seth argues in relation to the colonial context could to some extent be recalled within the present political frame of *democracy to learn*: 'colonial governmentality functioned to posit the possibility of self-governance and incite the desire for it, while simultaneously declaring it unachievable' (Seth 2007, 123). Democracy and autonomy are ultimately the yardsticks of the good citizen that postcolonial subjects cannot fully actualize.

DEMOCRACY AS A STRATEGY OF CONTAINMENT

Recalling Ranajit Guha's reflection on the nation-state as a 'strategy of containment' (Guha 2003), I suggest looking at democracy as a kind of *strategy of containment* and as a *discipline of mobility*. According to European analysts, development is the precondition for the settling of a 'safe' democracy and, conversely, projects of development require a minimum degree of social security in order to be able to really foster processes of democratization. In order to tackle these issues I have focused on how the International Organization for Migration (IOM) set its activities in Tunisia through two main strategies: democracy as containment and secured mobility through development. These two combining political technologies for governing revolutionized spaces centre on would-be migrants and returned migrants in order to mobilize wider moral economies of development. IOM functions as a norms-making agency, making space for new discursive and non-discursive practices of intervention aimed at governing mobility. The government of would-be and returned migrants has today a paramount relevance in the construction of a 'new democratic Tunisia', in which migration policies are less concerned with border management than with the migration-development nexus.

Responding to social unrest and the migration of thousands of young Tunisians towards Italy and France, IOM launched a campaign funded by the European Union addressed at 'stabilizing at-risk communities', namely, the inner regions of Tunisia where the revolution started and which are, at the same time, zones with a high percentage of emigration. This campaign was put into action supporting local investments and facilitating would-be migrants who planned to start small businesses. But the project also targeted migrants recently returned from Europe, staging for them 'reintegration paths': the logic that underpins these projects basically consists in making migrants learn to be responsible citizens in the face of the 'democratic challenge' and the historical revolutionary moment. In July 2012 I visited Tunisian villages of the inner regions of the country in order to find the developmental projects in support of returned migrants that IOM, the EU and the World Bank sponsored on their websites and in official documents. In fact, the overall picture that emerges from those governmental strategies is of the necessity of managing instabilities in order to build the 'new democratic Tunisia', to be supported on two main pillars: good governance and the rule of law. However, in those villages only a few people knew about the reintegration projects for returned migrants, which ultimately consisted of quite unskilled economic activities.[4] Included among the criteria for allocating funds is a true will to stay in Tunisia and not to leave again. However, no official data exists about effective departures and temporary migrations and returns from/to Libya: informal transports, constant homeward journeys and no register of entries and exits make the materiality of migration practices a multifarious phenomenon which cannot be fully readable and encoded by governmental programs. The complex texture of governmentality is constituted also by self-organized migration patterns and labour economies deployed in between the folds of the governmental machine. This suggests that in order to explore how a space is governmentalized we need to take into account practices of self-organization, political technologies and strategic embodiments of migration categories and identities. Despite the deployments of many European projects of development in revolutionized Tunisia, the script of a smooth transition to democracy fails to allow for the heterogeneous and complex reality of that space. This complex and ambiguous entanglement between the politics of mobility and of development shows clearly that migration governmentality largely oversteps the field of border policies, situating governing populations and would-be migrants within broader developmental technologies. In fact, all Tunisians coming from 'at-risk communities' are considered would-be migrants, due to the possibility of migrating opened up by the revolution. In the name of democracy as a stage to be fully achieved through transition, (some) would-be migrants are fixed in space: most people are not required to stay in one place, thus the politics of selected mobility that the European Union promotes for specific migrant categories—students, high-skilled workers—is refracted into a complementary politics of 'democratic containment' which encourages the building of durable economic perspectives in the country. Actually, since 2005 the IOM has put into place campaigns of 'sensibilization' aimed at convincing people not to leave their own country in an irregular way, showing the risks of clandestine departures and the uncertainty surrounding their future in the country of destination (Andrijasevic and Walters 2010; Georgi 2010). In the case of revolutionized Tunisia the geographical area where the revolution started is designated 'at-risk'; the goal is to prevent people from leaving the country, taming migrations and social unrest at the same time. The EU-IOM projects show quite clearly the ambivalent role that democracy plays in migration governmentality discourses concerning revolutionized spaces: 'the project provides alternatives to communities at risk, by promoting stability in a transition . . . addressing youth un- and underemployment in at-risk communities, through activities to enhance their employability in local and foreign labour markets, and to promote local socio-economic development' (EU-IOM 2011).

By encouraging people to foster local economies, both the EU and IOM prevent unskilled would-be migrants from leaving the country. In this way, safe-secured mobility is fostered by the conditions of development and economic growth: those who don't migrate avoid falling into the circuits of smuggling and illegal migration, while skilled migrants are encouraged to go to Europe to learn good practices of development. The first thing to remark about the development projects is that they enhance an economy of debt: in order to start the project, returned migrants need to ask for loans from microcredit institutes or ordinary banks. Secondly, those projects actually enforce a low-skilled economy: despite the injunction for a developed economy of growth, at closer scrutiny of enterprise activities supported by the European Union and the African Development Bank via IOM reveals that most pertain to the so-called economy of subsistence and to short-term projects just to prevent people from leaving (Cassarino 2012).

However, far from reading this in terms of an economic backwardness enhanced by international actors, I suggest drawing on Kalyal Sanyal's considerations, bringing attention to the constitutive heterogeneity of a capitalist economy and the simultaneity of processes creating, restoring and destroying traditional modes of productivity. As Sanyal notices, the basic needs-based approach to development is grounded on two main pillars: on the one hand, 'the purpose of developmental interventions is to create and extend entitlements outside the capitalist space for the excluded and the marginals' and, on the other hand, 'the employment strategy promoted by these organizations is highlighting the prospect of self-employment' (Sanyal 2007).

Such a perspective enables us to unpack the logic of transition to democracy: the politics of fixing (some) people in space and settling hybrid economies—private investments and unskilled activities—is also a strategy for taming revolutionary turmoil, maintaining revolutionized spaces in a state of permanent transition. In reintegration programs migrants are depicted as the counterpoint to the figure of the responsible citizen: returned migrants emerge, in this light, as subjects who irresponsibly fled the country in the aftermath of the revolution. However, a further moral partition is drawn among returned migrants themselves, splitting them between the responsible citizens and the incorrigible ones: indeed, those who were deported are excluded from any migratory or developmental program. In this sense, a sort of corrective reintegration pattern is envisaged for migrants who voluntarily returned to their country of origin, excluding those who came back as deported. In a nutshell, the rallying cry of democracy puts into place a normative matrix assessing which conducts are suitable for democratic standards. It is an effort in both economic and in moral terms: the attitude (disposition) to work hard and to rationally plan economic strategies is part of the same organizing principle. Therefore, moral development and economic development, however unachievable, are mutually entangled (Rose, O'Malley and Valverde 2006; Watts 2003). The European Union has established the future directions of revolutionary uprisings according to a secularist political teleology. Do the rallying cries of freedom and democracy have another connotation and meaning than in the European space? The secular and progressive narrative emerges in a quite outstanding way in the discourse pronounced by Cecilia Malmström, the European Commissioner for Home Affairs: 'I was impressed by the people's determination to make their liberated country a success. Here, and throughout the region, we need to constantly assess whether our policies are providing an effective response to their historic challenges'.[5]

This digression on the impossible autonomy of the teachable subject shows quite well that containment of mobility is precisely at the junction between the moral and spatial control of conducts on the one hand, and the government of populations, on the other. In some way, the hinge position between disciplinary power and biopolitics that, according to Foucault, is embodied by sexuality can be similarly ascertained in mobility, or at least in 'unruly' forms of mobility. If in the case of sexuality the articulation of disciplinary and biopolitical regulatory mechanisms takes place at the threshold between individual and population, in the government of mobility it is rather located at a spatial juncture between fixation to a place and regulation of movements on a transnational scale.

WORKING AT THE MARGINS OF POWER AND MIGRATION AS A 'DECOMPRESSION CHAMBER' OF THE REVOLUTION

Taking migration as an analytical standpoint for grasping the functioning of the control over lives involves situating the analysis at the margins of governmentality. By margins I mean the spaces and the subjects that function as elements of friction and resistance to governmentality. In this vein, Foucault's suggestion that to take the point of view of the reversal of and the limits of a certain configuration of power is essential to analysing its dispositive functioning (Foucault 1980a). These limits also designate the edges where many dispositifs of government overlap and where they need to reinvent strategies in order to respond to migration turbulence. In fact, as Deleuze and Guattari also put it, a dispositif of power is mostly characterized by what tends to escape from its grip (Deleuze and Guattari 2007). Taking the margins as an analytical standpoint means bringing to the surface mechanisms of governmentality that are particularly conspicuous at the borders, showing the uneven functioning of power—namely, how it works differently at the margins (Mohanty 2003).

The departure of young Tunisians engaging in *harraga*—literally, the act of burning frontiers—represents an apparently oblique phenomenon that actually complicates the narrative of the revolution and, at the same time, provides a view of the revolutionary events as uprisings and refusals against the government of lives. Revolts and turmoil produced slacker border controls, since many of the Tunisian *Garde Nationale* deserted or were employed to tame the riots in Tunisian cities; but at the same time the Tunisian government strategically seized the opportunity to send out of the country many youths who participated in the revolution. Indeed, in the face of an ungovernable mob, the strategy of the 'decompression chamber' appeared as a temporary solution, and the spontaneous practices of migrations that the revolutionary *elan* mobilized ultimately were facilitated by not obstructing departures by boat from the cities of Sfax and Zarzis. In this sense, paradoxically, the primary connection between revolutionary turmoil and practices of migration was established by the Tunisian government, who depicted Tunisian migrants as the sons of the revolution: young people who might provoke disturbances and the snowballing of social and political unrest should literally be encouraged to put out to sea. Tunisian migrants who left the country after the outbreak of the revolution resembled a 'migratory mob', namely, an ungovernable social problem whose departure it was best to facilitate.

THE UNCEASING PRODUCTIVITY OF SPACES

Democracy as a strategy of containment has to be seen as a complex response to practices of mobility which unsettled the normativity of previous spatial economies. At close scrutiny, what comes out is the reconfiguration of existing spatialities along with the production of new bordered spaces and special zones that in turn multiply borders and frontiers: the differentiated access to borders and spaces is enacted precisely through the tracing out of temporary zones, special economic areas, regional cooperation and highly monitored spaces. For instance, in the case of Neighbourhood Policies a shared political Mediterranean space is envisaged in the name of the proximity of the two shores of the Mediterranean. But such a commonality is actually based on an asymmetry, as is clearly shown by the principle of conditionality which underpins the logic of neighbourhood (Cuttitta 2010; Hailbronner 1993; Lavenex 2008): the 'migration clause' is at the core of broader economic bilateral agreements, compelling North African countries to adopt measures of reinforced border controls against migrants' departures and to accept the repatriation of third-country nationals on their soil. Another important reference for the multiplication of special zones produced by migration policies is the Regional Protection Programs based on the principle of the redistribution of the 'refugee burden'; these provide economic and political incentives for non-European countries to adopt a politics of asylum, in order to strand asylum seekers before arriving in Europe. After all, the politics of asylum is also functioning in part as a politics of containment *by stages* and *through channels*. The leading logic is to predispose external spaces of protection, encouraging partner states to activate systems of humanitarian assistance with the supervision of European agencies. The second pillar of this package is to create *safe environments* in countries of origin in order to make displaced people 'voluntarily' return there. Finally, the Regional Protection Programs also include the resettlement project aimed at resettling refugees stranded in third countries without asylum policies: special zones and regional externalized spaces of protection overlap with humanitarian zones breached within the European space. However, if, on the one hand, it is important to critically stress the ongoing trend to externalize the mechanisms of asylum, on the other, it is important to disengage from a Eurocentric vantage point, exploring whether and how, from their side, third countries could gain autonomy in setting a politics of asylum on their territory.[6] In particular, the possibility of building an economy of asylum independent from Europe could challenge the dominant discourse on international protection which posits European democracies as the real model guaranteeing an efficient and abiding system of asylum.

This last point leads us to problematize the limits of a critique of the border regime conducted from the northern shore. How is it possible to resist European migration policies and bilateral agreements without gaining (relative) economic autonomy from the European labour market? What are the margins for a politics of externalization that could be strategically used by North African countries, reversing the migration clause to their own advantage? In Tunisia the debate on relatively autonomous development, independent from the European principle of conditionality, is considered a precondition for thinking about democracy and improving participative politics; and in this sense it could be contended that the Arab uprisings troubled the Eurocentric left-wing critique of development. Meanwhile, three years after the outbreak of the Tunisian revolution, the high rate of unemployment continues to generate social protests throughout the country and movements demanding the cancellation of the national debt. Thus, if migration is always encroached upon and encrusted in other domains, and must be situated within the broad government of conducts—as a government of the mob—a critical analysis of migrations should rethink together mobility, economic and political struggles. Migration could hardly be assumed as an autonomous subject of analysis or as a pure phenomenon: as a catchword and as a site of struggles, it is the catalyst where a huge variety of economic and social issues converge and overlap. The marginality that the migratory issue has in the Tunisian official political debate is quite astonishing: in fact, after the huge wave of migrations towards Lampedusa in early 2011, the migration topic then faded again into the background, promptly reactivated in the light of shipwrecks of migrants' boats or via the debate on the economic crisis, unemployment and projects of development. However, moving away from the institutional channels, in Tunisia migration is perceived instead as a socially rooted phenomenon: 'at some point (many) young people decide to leave, to do the harraga, sometimes with the desire of living abroad for a period of time, and sometimes looking for a job. There is nothing new in this, it has been happening for ages'.[7]

A debate on democracy has long existed in the revolutionized Arab countries and did not suddenly mushroom in 2011 as a follow-up to the

'Arab awakening'. Yet, if we follow the European narrative, the practice of democracy in the Arab states came out of nowhere and in some way went against the 'natural disposition' of the Arab people, historically prone to resist social transformations (Pollack 2011; Bishara 2012). Nevertheless, the point is not to engage in a quest for the original meaning of democracy in the Arab context, retaining a supposed authenticity of the signifier, but rather to highlight the complex affiliations and inflections of that political referent: the way in which it is moulded by colonial heritage, the resistance to the importation of the values of the colonizer and a reworked meaning of democracy itself (Filali-Ansary 2012; Ramadan 2012). This is crucial for showing that the 'battle over democracy', both as a resistance against the colonial power and as a debate about the construction of a new society, has a long historical trajectory. In other words, it makes it possible to trace back the tale of the 'Arab Spring' to its colonial legacies without assuming the revolutionary uprisings of 2011 as a sudden awakening of the region. In fact, as Massad reminds us, 'there have been a number of uprisings in the past several decades in the Arab world, foremost among them the Egyptian uprising in January 1977 against President Anwar el-Sadat's austerity measures, which Sadat dubbed an "uprising of thieves" and the West dubbed "bread riots"; the Sudanese uprising against the US-backed dictator Ja'far Numeiri in 1985; and the Palestinian uprisings of 1987–93 and 2000–2004 against Israeli military occupation—none of which merited the term Spring' (Massad 2014). Moreover, such a move suggests undertaking a spatial dislocation from Europe as a unique standpoint of analysis and main epistemic referent for analysing the signification of political languages and categories that sometimes have the same name.

The ambivalence of the 'democratic conquest' as mapped by the European gaze is translated by migration and governmental agencies into the idea of *complex crisis*. This expression conveys the difficulty in reading those upheavals and making them intelligible to ordinary political narratives; and at the same time it postulates the necessity of framing a global political approach to that turmoil, encapsulating migrations and social conflicts within the category of a socio-political crisis. Consequently, migrations turn out to be a part of a broader spatial re-assemblage which is presented by governmental actors and migration agencies as a 'coherent approach in the area of migration, mobility and security'.[8] The reference to the complexity of the Mediterranean crisis and the claim for a comprehensive approach to it should be read as a way for coming to grips with political space in motion—the space of the Arab uprisings—that is slipping out of the hands of governmental narratives. In the following section I examine the Mediterranean crisis, focusing in particular on the way in which the catchword of crisis has been mobilized by migration agencies and governments in order to construct packing the *twofold spatial upheaval* as a new space/object of government.

MIGRATION (IN) CRISIS AND 'PEOPLE NOT OF OUR CONCERN'

The Slippages of the Migration Crisis

The two-year period 2011–2012 could be seen as the age of a 'Mediterranean crisis' spanning from the edges of Africa—the Libyan war—to the countries of southern Europe. This analysis centres on the Mediterranean, seen in many political analyses as the space in which two crises overlap: the 'migration crisis' and the resulting economic backlash. As I explained in the beginning of the chapter, the migration crisis is figured as a flood coming from the southern shore of the Mediterranean: the spatial and political upheavals produced by the so called Arab Spring and brandished as the democratic awakening of the Arab countries were very soon stigmatized as a social turmoil and as a migratory disorder, mostly when the long-awaited bank effect on the northern shore actualized in the presence of thousands of migrants on European soil. It is by following the instabilities produced by the economic and migration crisis that I orient this reflection. Focusing on the politics of mobility, migration agencies have disconnected the relationship between the European governmental crisis and the crisis of the asylum system—positing the latter as a humanitarian and security problem. At the same time, they have seized the crisis as a floating signifier for setting up 'migration in crisis' as an odd compound where different kinds and meanings of 'crisis' collapse into one another and become conflated: the crisis in Libya, the humanitarian crisis at the Tunisian border, the crisis of the European states receiving thousands of migrants and finally the crisis of the migrants who have arrived in Europe and are now struggling with the economic recession.

The script of a 'migration in crisis' has been recently promoted by IOM in order to address the Libyan political turmoil and its disseminations, namely, its impacts on other spaces—Tunisia and Europe—and in different domains such as the humanitarian regime, security and economics. It is certainly not the first time that the paradigm of the crisis has been introduced by states or international agencies as a keyword encapsulating an array of political technologies of migration governance.[9] But in the aftermath of the Libyan war, the catch-word of the 'crisis' has been reintroduced as a multifunctional prism for framing a heterogeneous array of 'mobility disorders'—namely, practices of migration that through their 'spatial takeover' (Sossi 2012) trouble (b)ordering spaces.[10] But what does 'migration crisis' stand for? It is not merely quibbling with words if we take into account the nuances of that formula, especially the swing between 'migration crisis' and 'migration *in* crisis'. In fact, the use of the two expressions reveals a slippage in the meaning of the crisis when referring to migration: 'a large scale, complex migration flow resulting from crisis and typically involving significant vulnerabilities for the individuals and communities affected . . . migration caught in crisis involves different categories' (IOM, 2012b). It is noticeable that the crisis refers to the state of precariousness, vulnerability and restricted mobility which affects migrants crossing the borders of a third country due to a military conflict and people who became 'migrants' because of the crisis; and at the same time, it addresses the economic backlash and the security issue affecting receiving countries. Migration as a 'disordered practice of mobility' is staged as a turbulent and troubling factor in itself, triggering a state of crisis or fostering an ongoing crisis already present, irrespective of the nature of the crisis—humanitarian, economic or security. What the regenerated formula of 'migration crisis' makes visible is that migration works precisely as a magic buzzword through which the conquests of democracy and freedom have suddenly been translated into an unfulfilled democratic revolution: 'migration' plays as a transformative catalyst for re-codifying political struggles and spatial upheavals as sources of undetermined crisis: migration as the space-troubling factor, migration as the degenerative force of a favourable mobility, migration as the deviation from the road to democracy, migration as a plight for social cohesion and as a disobedient practice of movement.

A Parenthesis on the Blurred Catchword of 'Crisis'

In his genealogy of the variegated occurrences in which the term 'crisis' has historically been used, Reinhart Koselleck shows the catchword function of the notion of crisis and its blurred meaning, which covers a wide semantic range drawing on multiple domains—medicine, law, theology, philosophy of history—retracing the emergence and the uses of the term 'crisis', the political-economic signification emerges only in the late eighteenth century and still continues to encroach upon other domains (Koselleck 2012). Nevertheless, the political force of this notion is historically grounded on its plurivocal nature, evoking different and overlapping levels of meaning. From the second half of the nineteenth century, 'crisis' has become one of the main keywords of the political vocabulary for legitimizing structural reforms as well as for marking points of no return or critical moments of transition which impose a quick response or mutually exclusive choices. Ultimately, what relates the migratory issue to the notion of crisis from a theoretical and political standpoint is the hybrid nature, namely, 'the qualities of creating connections and at the same time the necessity to connect itself to other terms' (Koselleck 2012, 92).[11]

The Tunisian Migration Cluster

Despite the formula 'migration crisis' being coined to address the Libyan political turmoil and its multiplicative effects, revolutionized Tunisia is an interesting space for interrogating the formation of what I would call a 'migration cluster': after the outbreak of the Tunisian revolution and of the Libyan conflict, Tunisia has become a space of 'complex' migrations, as a factory and at the same time a recipient of migrants and would-be migrants, sub-Saharan refugees and asylum seekers, young 'Tunisian beggars' crossing the Mediterranean and Libyan nationals. But Tunisian citizens who left the country for Europe and the migrants who fled Libya were seen in Tunisia as two completely different phenomena. This consideration is in part true if we consider the different conditions and reasons for migrating, but at the same time it overshadows the commonalities that depend on the very mechanism of 'selected mobility' from which both these practices of migrations are excluded. In this regard, the existence of the refugee camp of Choucha and the crossing of the Libyan border by almost one million people in 2011 after the outbreak of the war have remained in the shadows in comparison to other political issues taking place in revolutionized Tunisia. Or rather, the problem of migrants and refugees fleeing Libya was primarily tackled by the 'popular chain' of hospitality set in place by the Tunisian people (Tazzioli 2012).

'Give Us Our Lives Back': Stranded Migrants Out of Place in the Space-Frontier of Choucha

Choucha is a tent camp in the middle of the Tunisian desert, nine kilometres from the Libyan frontier of Ras-Jadir and more than ten kilometres from the closest Tunisian village, Ben Guerdane. And since the problem of the rejected refugees and of the not-resettled refugees arose—in summer 2012—the stranded people, waiting for months as in a lottery, have remained in the shadow of the Tunisian political debate centred on the construction of a 'new democratic Tunisia'. The camp opened on 26 February 2011, hosting people displaced by the Libyan war, those hundreds of thousands of Libyan residents—almost all of them 'third-country nationals'—who fled the conflict to Tunisia. The maximum number of people trapped in that space peaked at twenty-two thousand between March and April 2011; the following year the average number of asylum seekers stranded there was around four thousand, and by March 2013 an estimated nine hundred eighty people were still there, although the exact number, as I will explain later, was very difficult to establish.

From the outset the main sorting at Choucha among war-displaced people was between those who decided to return to their country of origin and those who applied for asylum; then, when UNHCR started to communicate the results of the demands for asylum in 2012, the camp was soon split into two areas: the official camp and the areas of the rejected refugees. The latter were invited to leave Choucha, as they were out of place, since UNHCR takes into account refugees and asylum seekers but not those who 'failed the trial' of the asylum process. However, the Libyan crisis and the claims of the would-be refugees brought into question the tenability of the very principles of the international protection determined by the Geneva Convention and the disregard for the reality of the international labour regime which causes people to move all over the world: the outburst of the war produced an unprecedentedly huge outflow of third-country nationals in Egypt and in Tunisia due to the presence of more than one and a half million migrant workers in Libya.

The criteria for recognizing the status of asylum materialized in the interview with the UNHCR Commission, in which people were asked why they left their country of origin, instead of being asked the reasons why they escaped Libya, which is their country of residence or in any case the country where they had worked for years. 'All of us fled from Libya, from a war' rejected refugees stated during their sit-in of protest in Tunis, 'and so no distinction should be made among us, between those deserving of protection and those who do not'.[12] If in the first period rejected refugees were sheltered and assisted by UNHCR, despite the spatial seclusion, since October 2012 no food and medical assistance has been provided to them.[13] Moreover, those who were employed by NGOs in the camp were dismissed from work, so the only possibility of getting food now comes from finding an informal job in the village of Ben Guerdane. Confronted by the protests of the rejected refugees and the political denunciation of some local and international groups of activists, UNHCR reiterated that 'rejected refugees are not people of our concern, so we are by no means obliged to take care of them' adding that 'in a time of crisis, we have to cut the costs for managing the camp and the assistance provided to denied refugees until October was not owed'. 'They are people not of our concern' stands for 'these existences are not visible for us, and even less are they of importance to us; to someone else must go the task of governing them'. In fact, rejected refugees exist precisely because by labelling some as 'rejected refugees', the mechanism of asylum de facto creates 'illegal' migrants: 'We do not say that denied refugees are illegal migrants. This is a question out of our domain of concern: we simply say that they cannot be protected under the criteria of the asylum system. Thus, we can say who they are *not*, while it's up to the competence of nation-states to decide upon their juridical status'.[14] In part, this is a strategy designed to deter those unplaceable subjects. Nevertheless, the production of people 'not of concern' goes along with the control over their mobility. In the case of Choucha, the passports of the rejected refugees are still in the hands of UNHCR, which releases the documents only provided the person returns to their country of origin: a unilateral pact is thus set in place by the humanitarian regime, which locates an exchange of security-mobility at its core. Would-be refugees are allowed to become undetected presences in the Tunisian space, and UNHCR encourages them 'to find a job here, to value one's own skills as economic migrant'.[15] Therefore, the so-called migration regime is actually composed of and fragmented into different and sometimes conflicting governmental agencies, research centres and states, each one with very specific tasks and domains of concern, which produce subjectivities that largely exceed their domain, despite UNHCR's formulation that 'they are not people of our concern'.

The Crisis of What? The Free Spinning of the Sorting Mechanism of Migration Governmentality

This snapshot on Choucha has unfolded the slippages and ambivalences at stake in the use of the formula 'migration (in) crisis': the crisis refers to the potential turmoil and the demand for resettlement in Europe in the face of the presence of would-be refugees on Tunisian soil, but at the same time it addresses the condition of being caught in crisis as migrants, denied refugees, asylum seekers, un-resettled refugees and a plethora of 'troubling' forms of mobility. Nevertheless, I'm not suggesting that these impasses have been generated by the Libyan crisis. In the end, what I have mentioned here are not exceptions to the functioning of the government of the humanitarian, and it is not in terms of violations of the rules of asylum that we can sort out a *migration population*—the separating of refugees, denied refugees, economic migrants, etc.—and consequently exclude some of them from the regime of protection. In fact, the way in which refugees and migrants have been classified into migration profiles is not too dissimilar from other contexts: UNHCR produced in Choucha an array of degrees of unprotection among which the denied are obviously those with no place at all, 'a nowhere as their condition of existence' (Sossi, 2007). Rather, the Libyan crisis has finally exploded the tenability of using the country of origin as the basis for the logic of asylum; and this was made visible by the rejected refugees who carried on protests in the name of their common (forced) escape from Libya. They put into place a 'politics of the governed' (Chatterjee, 2004), stressing the fact that they were all being subject to the mechanisms of migration governance and, at the same time, escaping the Libyan conflict. In fact, in contrast to the mechanism of *partage* and to the country-based criteria (people coming from 'safe' or 'unsafe' countries), rejected refugees demanded resettlement away from Tunisia. To the primacy of the safe/unsafe list and of national origins, rejected refugees impose the law of their spatial presence and of their condition of being governed by the 'migration bouncing game',[16] which strands people by fixing them in spaces or making them wander without a place. Moreover, they reversed the very logic of (un)safety, demanding 'to be resettled into a safe country', thus excluding the possibility of remaining in Tunisia with a temporary protection status, as the UNHCR was advocating.

The Libyan crisis and revolutionized Tunisia also make visible how the economic crisis, the political upheavals in the Arab countries and the epistemic crisis of migration categories sharpened and transformed the mechanisms of migration government. The multiple crises exploded the unquestioned functioning of the asylum system's partitioning machine and, at some points, some of those mechanisms spun freely. But while this temporary short-circuiting of the system went to the advantage of some of them—through the concession of a temporary permit in the case of the Tunisian migrants arriving in Italy in 2011—the *epistemic crisis* of governmental migration agencies and the juridical confusion produced by 'complex migrations' (IOM 2012d) were detrimental to the would-be refugees in Tunisia, who were literally stuck in juridical impasses. Indeed, the migration crisis is recognized internationally as also a crisis in migration governmentality, an epistemic crisis within the mechanism that sorts people, makes up juridical subjectivities and spatializes their conducts: 'the difficulty . . . is that a complexity of mobility practices is also at play: mixed migration flows formed of people moving for diverse reasons and with different aims, generate challenge to migration management' (IOM 2012c). Thus, the economic crisis which dramatically impacts on migrants' lives and further tightens the already restrictive European politics of asylum and resettlement is coupled with the crisis of the 'migration sorting mechanism'. But at closer scrutiny the 'partitioning log jam' of the migration regime in classifying people into mobility profiles is, as governmental agencies recognize, the outcome of migration upheavals: practices of mobility which to some degrees exceed existing partitioning criteria and cannot fit into those migration profiles. What the governmental lexicon calls 'complex migration', resonating in the designation 'mixed migration flows', corresponds to the juridical confusion generated by migration practices that are not 'expected' and whose combination of status, citizenship and country of residence makes it difficult to trace uncontested juridical profiles.

The Moral Economy of Resettlement and the Secrecy of Humanitarian Knowledge

The complex mechanism of resettlement as a technology for governing the migration population needs to be situated within a broader moral economy of states in a time of crisis. A system of economic incentives has been activated by the European Union which grants up to 6,000 euros per refugee to states that resettle refugees; and countries like Brazil enter the program in order to promote themselves as democratic states on the world scene. However, the criteria that countries adopt for selecting and excluding people remains secret. Actually, we might guess that skilled migrants are the most desired refugees, and that logic would corroborate the governmental discourse which in principle promotes skilled mobility.

Going in-depth into the mechanism of resettlement, however, reveals that this criterion of selection is not really the principal one. As the refugees in Choucha have understood while they were waiting for their turn, Canada accepts only francophone people; Portugal tends to take those refused by other countries; Denmark takes the vulnerable cases; and Sweden and Norway prefer women. Among the recognized refugees in Choucha, around 150 people have been labelled as not eligible for the resettlement program: people with legal precedents, people taking part in social disturbances and people charged with terrorism. Following the pre-selective screening carried out by UNCHR, nation-states interested in resettling people select the most suitable profiles among the refugees and refuse many of them on the grounds of 'security reasons'. In this way, in Choucha some of the official refugees won't be resettled, despite their juridical status. At the moment of writing, the unelected refugees come from Arabic countries, Palestine and Iraq.

The knowledge possessed by would-be refugees at the Choucha camp involves a kind of lateral thinking here: perfectly aware of being stranded on the international chessboard of the politics of mobility, they are nevertheless able to work out their future location via a process of deduction, observing how, despite the formal criteria, the divvying up of people is really done. What is supposed to constitute a shared and standardized process, then, under close scrutiny turns out to be a mechanism of exclusionary knowledge. And just being wise to their rights and the formal procedures of the asylum system cannot be of help in this case: despite their deep knowledge of international law on protection and of the geopolitical context, the would-be refugees realized that a huge discrepancy persists between the order of norms and the effective government of their lives. Who holds their passports, into what migration profile they have been fit by states, where their dossiers have been placed and what is on the list of safe/unsafe countries: all this information is unknown to the refugees. The list of safe and unsafe countries was established by UNHCR in the late 1980s with the purpose of pushing through the procedure of asylum (Hailbronner 1993).

Would-be refugees fleeing Libya have to some degree poured the crisis into the logic of protection. The unexpected arrival of hundreds of thousands of third-country nationals in Tunisia, and the rejected refugees' demand that international protection be afforded to everybody fleeing Libya de facto undermined the tenability of the very logic of the asylum, which relies on a partitioning rationale—economic migrants/refugees, bogus refugee/vulnerable subjects, denied asylum seeker/resettled person. Confronted with the incorrigibility of their demand, UNHCR has adopted a tactic of discharge, producing rejected refugees who actually can neither stay in any place nor move anywhere except back to their country of origin. Therefore, they become undetectable presences in the Tunisian space. The elusiveness of the numbers regarding the rejected refugees in the camp is due to humanitarian actors' tactic of chasing people away, which pushes some people to abandon the camp to get informal jobs or go back to Libya. To discharge the many in order to care for a few is not a novelty coming out of the 'migration (in) crisis' script but is rather the underlying logic of the international politics of asylum. However, as I stressed above, this rationale becomes more tangible in a space of crisis. Meanwhile, in the Libyan case the migrants' spatial upheaval made the mechanisms of partitioning spin freely: the 'migration (in) crisis', unlike other historical times of crisis, has not worked as a moment for radically reassembling power relations or for transition to another regime of government. Rather, the spatial and political disorder, along with the multiplication of *precarious spaces*—such as zone of humanitarian crisis—have been played as a means and object of government (Sidaway 2007).

The Politics of Presence: The Refusal to Stay in One's Own Place by Taking One's Own Space

This gaze on the Choucha camp highlights the limits of a hyper-governmentality grid for analysing the politics of mobility and the mechanism of protection. In fact, the tactic of 'discharge' largely prevails over the logic of managing the lives of all would-be refugees; while, at the same time, the mobility and conduct is highly monitored. In particular, following are two episodes which show very clearly would-be refugees' conditioned and monitored (im)mobility.

Ben Guerdane, 26 March 2013: on their way to the World Social Forum that took place in Tunis, a group of refugees from Choucha camp travelling on three buses were blocked at Ben Guerdane by the Tunisian national police. 'You are not allowed to circulate in Tunisian territory' the policemen argued, disregarding the special permit that the refugees had obtained from the Defence Ministry to go to the Forum. They had seized the opportunity of the Social Forum to make their voices heard, as the name of their blog—'Voice of Choucha'—also suggests, and to demand that UNHCR 'finish its work' by acknowledging their status as Libyan war refugees and resettling them in safe countries. After their confrontation with the police, eight of them succeeded in reaching the Tunisian capital by a collective taxi, sneaking away from the police blockade. The next day, only half of the people who had been stopped by the Tunisian police would manage to arrive at the Forum.

February 2014: a group of refugees still living in Choucha set up a protest in front of UNHCR headquarters in Tunis. All of them were arrested and put into jail: the condition for being liberated was to return to Choucha and not protest anymore, accepting UNHCR's local integration program. The refugees refused the deal and decided to remain in prison, arguing that ultimately the jail is a more liveable place than Choucha and that they would never give up their struggle for finding a space to live. After twenty days they were released and carried back to Choucha, in the desert.

The decisions of these rejected refugees and non-resettled refugees to stage these protests based on their different juridical status draws our attention to the ambivalence surrounding the issue of pluralizing and differentiating migrations. In fact, while, on the one hand, the splitting of the Choucha refugee group was the result of their strategic consideration that two different demands should be addressed to UNHCR, on the other hand, such a decision highlights the appropriation by refugees of the partitioning categories of the asylum, thus limiting any possible broad alliance or common ground of struggle among migrants. In a nutshell, the epistemology of the humanitarian regime is predicated on the multiplication of mobility profiles, which ultimately fragments migrant struggles. Thus the important task of pluralizing the migration, stressing the heterogeneity of migrants' conditions and of their stories, should however take into consideration the migration agencies' strategies of fragmenting by differentiation. In the face of that, the concern is to stage the multiplicity of migrants' conditions without dividing and weakening possible common struggles of migrants stranded in the same space.

December 2012–August 2013: 'They are not people of our concern, anymore', repeated the UNHCR officer in Zarzis, 'so it's not our problem what they do with their lives. They are not vulnerable or at risk. It's their life; we are not responsible for them and it's not our fault if they die going to Italy by boat'.[17] From this statement and from a knowledge of the rules of the game of the asylum system it seems clear that there was no room for further developments and reopening of the dossiers of the rejected refugees, and that only a non-institutional solution could be envisaged, outside of the formal recognition of protection. The logic of the rejected refugees was quite different, however: 'Since they govern us, they must take care of us. And they need to comply with the principles and the work they are expected to do: 'they should take decisions based upon the rights they talk about—"human rights"—but "humanitarian forces that in principle should defend our rights mock us and strip us of those rights".'[18] Therefore, the governors were seen as the truly 'bogus' ones, to be opposed and unmasked: 'We know that our condition is an international affair, we are part of an international problem which concerns also Palestine and Iraq. So we don't leave the camp, we do not accept their game and we stay here as long as the work of UNHCR remains unfinished'.[19] The occupation of the camp continued until 30 June 2013, when the camp was expected to close, but the rejected refugees resisted eviction, demanding that a solution be found for them. Most of the rejected refugees have chosen to stay at Choucha, imposing the law of their presence: 'They cannot but see us, they want to make us invisible, but we are here'.[20]

The insistence of the rejected refugees that they be afforded a form of protection, despite being declared by UNHCR as out of the 'rules of the game', amounts to a quite peculiar way of endorsing the condition of 'being governed': on the one hand, they can only play within an institutional horizon, claiming to be recognized as refugees, appropriating and making use of the same vocabulary and the same partitioning categories of migration governmentality; but, on the other hand, they appropriated and hijacked the rules of the asylum game, pushing for an impossible demand—'protection for all'. As a matter of fact, that demand was 'impossible' and also paradoxical within the rules of the game of asylum, which, in the end, both activists and critical researchers take for granted in thinking about political strategies. By occupying the camp, despite their condition as subjects 'out of place', rejected refugees in Choucha refused to stay in their own place (Fanon, 2007), the paradoxical place of being without a 'legitimate' space.

Choucha beyond the Camp

The specificity of the 'humanitarian grasp' on migrants' lives in the camp is demonstrated by its bordered spatiality and by the 'temporal captivity' to which migrants are subjected, namely, their condition of indefinite strandedness. In turn, this tends to shape the camp as an experimental space for governing both migration populations' and individuals' conducts. However, far from being exceptional places governed solely by arbitrary rules and reducing the migrants to bare life, camps are sites where migrants' indefinite waiting combines with the production of juridical subjectivities of rejected subjects—the denied refugees—and, more broadly, with an activity of constant monitoring. Moreover, in terms of political technology and governmentality, 'more than spaces of exception, the different centres for asylum seekers are better grasped through the definition of "margin" as a space in which different forms of power coexist, whose goal is not necessarily the "readability" of the subjects; [in the camps] the practices of the implementation of the law come into conflict with the application of other forms of internal regulations' (Sorgoni 2011).

Therefore, 'Choucha beyond the camp' refers first of all to a move beyond the paradigm of the exception. It also concerns the spatial effects of the camp governmentality beyond its official spatial boundaries and at a temporal distance. Indeed the presence and the functioning of the camp exceeds the space of Choucha: the presence of humanitarian actors managing the life at the camp spread across different Tunisian locations (Zarzis, Medenine, the Ras-Jadir border) and on a daily basis many refugees move to the closest towns (e.g., Ben Guerdane) to do informal jobs or remain in the premises of the camp to get food and water—since, at some point, UNHCR stopped giving food and water to the rejected refugees. In particular, after UNHCR's official closure of the camp, Choucha has actually become the centripetal site of refugees' erratic movements. In fact, many of those who have been denied international protection or who have not been resettled in other countries have moved to Medenine or Tunis so that they do not have to live in tents in the middle of the desert. Some, however, frequently return to Choucha due to earning a salary insufficient to pay rent or because they are evicted from occupied buildings in Tunis. Furthermore, Choucha is now the main site where non-Tunisian migrants in Tunisia can obtain contacts for leaving Libya by boat. Finally, since the closure of the camp, Choucha has operated as a 'transit station' for migrants who escape Libya, although this phenomenon is ignored by humanitarian agencies, because, for UNHCR, the camp officially no longer exists.

From a spatial perspective it could be argued that the 'Choucha effect' has impacted a huge geographical area and, more precisely, that the militarized Libyan frontier has 'stretched' up to the camp. Indeed, the space of Choucha has become a military zone, constantly monitored by the Tunisian army, and a place where the control of migrants' movements overlaps with the fight against weapons trafficking that takes place between Libyan and Tunisian towns. Moreover, despite the camp never being at the core of the Tunisian political debate, it has been a crucial matter of negotiation between the European Union and UNHCR on one side and the Tunisian government on the other for pushing Tunisia to develop an asylum system.

As mentioned previously, the impact of Choucha 'beyond the camp' contains a temporal dimension, evident in the impact of the camp's governmentality over migrants' lives and their future movements. This is due to the juridical status that migrants receive in the camp as well as the experience of erratic mobility, an experience shared with many others, from Libya to Choucha, and up to Italy or northern Europe. This shared erratic mobility consists of border zones of waiting—like the months spent stranded at Choucha—and moments of acceleration or risk, such as crossing the Mediterranean by boat.

So, where is the previous 'Choucha population' three years after the opening of the camp? Singular stories and choices and the consequences of UNHCR decisions (that denied some of the migrants and recognized others as refugees) have certainly contributed to spread those migrants across the two shores of the Mediterranean: some work in Tunis, others are still in Choucha demanding to be resettled, others have died at sea and some are in different European towns. Governmental migration maps and images tend to focus on border crossings, 'exceptional' spatial sites of detention and hosting, or emergency contexts, but between those moments of visibility it is important to shed light on how borders impact on people's lives at a temporal distance. By that I mean two different but complementary things: on the one hand, it is important to know what happened to those migrants after Choucha, how their juridical status has influenced their movement; on the other, as I showed in chapter 3, migrants have collectively assumed their flight from Libya and the period in Choucha as two criteria for claiming international protection for all those who escaped the Libyan conflict. This ambivalent 'Choucha effect' brings out the nuances of the production of migrant subjectivities in the camp: in part the divisions between rejected refugees and refugees approved by UNHCR, and the other many migration profiles that it generates, have a considerable impact on migrants' future patterns and possibilities; but people also escape and struggle with those migration profiles. The juridical labelling that takes place is in fact only one of the subjectivizing[21] conditions (along with racial and gender issues, as well as labour conditions) that shape and impact on migrants in the camp and on their concrete field of possibilities for moving on.

THE TRANSNATIONAL MAGHREB AREA OF MOBILITY IN THE MAKING

From the perspective of an interrogation on the spatial economies that the Arab uprisings generated and transformed, it should be asked if those spatial upheavals have unsettled the national frame. For this reason, in order to challenge the methodological nationalism that usually underpins political analyses on mobility I turn the attention to the spatial outcomes that the twofold upheaval generated at a transnational level. The Maghreb space of free circulation is neither properly a new zone nor to this day a real space; however, the Arab uprisings have pushed for the enactment of an area of free mobility despite running into very conflicting positions. In fact, the Maghreb area of free circulation was officially established in 1989, but then it actually worked as a space of free circulation of goods and not of people.[22] Or more precisely, the facilitations for internal migrations within the geographical area including Mauritania, Morocco, Tunisia, Algeria and Libya have mostly depended on bilateral agreements (Perrin 2008). The reason why this space in the making is of particular interest for this analysis on migration is that it intersects migrations across the Maghreb region and human mobility, on the one hand, and economic spaces and political economy on the other: the free mobility of people and the making up of an economic space beyond the partnerships with the European Union need to be analysed together. In addition, the transnational space in the making does not correspond to the area in which the Arab revolutions took place; rather, it traces a potential new cartography of the entire Maghreb region as one of the spatial and political upshots of the uprisings. Indeed, neither in Morocco nor in Mauritania did people take to the streets in 2011, and in Algeria, despite protests against the regime for the rising price of bread in January and

February 2011, the political regime was not overturned. The project of a Maghreb area of free mobility needs to be situated into the broader African debate about the politics of mobility: what is criticized by many African scholars and analysts is the lack of a common African regime of asylum and mobility, as well as the incapacity and incompliance of states in coping with inter-African migrations, setting, for instance, a labour politics of free mobility between those states (Likibi 2010). In particular, they raise the need to establish concrete conditions for a politics of mobility independent of bilateral agreements with the European Union. In fact, as Romoulad Likibi questions, 'according to the present conditions, is it possible to talk of political partnership in the full sense of the term?' arguing that in order to negotiate with the European interlocutors 'no important decision about the future of the African continent should be left to others' (Likibi 2010, 146).

It is precisely at the crossroads of these two aspects that some pressing concerns arise around the migration and development nexus: To what extent should a quest for no borders and free mobility take into account radical changes in political economy in order to not be merely a liberal rallying cry? And conversely, how can a new economic space generate a space of mobility not based on the logic of markets? In this way, the focus on this contested-space-in-the-making allows us to problematize a discourse on free mobility that does not complicate the political analysis on borders and mobility with the economic issue: in particular, those perspectives reveal their limits in the light of political uprisings which made visible the fact that 'the centre cannot hold' (Dabashi 2012), namely, that Europe and the West are no longer tenable as epistemic and political referents. Neither the vocabulary of the uprisings nor the patterns of current migrations from North Africa view the European space as a dreamland, especially after the outbreak of the economic crisis which contributed to the dismissal of the desirability of Europe. Moreover, migration policies in the Maghreb region need to be read in connection with a spatial economy of people's mobility that goes largely beyond migratory patterns of going to Europe. Indeed, emigration from the Maghreb area to Libya and to the Gulf States is a quite longstanding phenomenon—that can be traced back to the early 1970s at the time of the oil crisis—and since 2008 it has further increased due to the economic crisis, displacing the balance towards the east (Paggi 2014).[23]

Thus, the Arab uprisings and the spatial-economic outcomes they engendered lead us to recast the relationships between a radical critique of borders and a reflection on alternative economy. But before unfolding this point, I refocus on the Maghreb contested space of free mobility. In January 2012 Tunisia proposed to the other countries of the Maghreb area to build a common space of free circulation not only for goods but also for the citizens of those countries. The political debate around this project wavered between two ambivalent orientations: the possibility of setting a space of economic cooperation, independent from the agreements with the European Union, and at the same time the goal of fostering the circuits of free circulation of people and capital to entice investors from abroad. The leading opponent of this proposal was Algeria, which cautioned against the possible illegal trafficking of weapons that such a measure might facilitate and warned of the risk of losing national identities. Nevertheless, that space of free mobility in part already exists, due to bilateral agreements between the countries of the Maghreb region. For instance, this is the case of the agreement between Libya and Tunisia, signed for the first time in 1974, allowing both Tunisians and Libyans to freely move, work and stay in the two countries. It is noticeable that the agreement was signed just in the aftermath of the oil crisis of 1973, which marked a considerable turn of Tunisian migration routes from France to Libya, since the latter experienced a period of economic growth and consequently of labour force demand due to the increase in oil prices. And while the Arab revolutions have been an opportunity for some states to revise these treaties, informal flows of people, most of all between Tunisia and Libya, have never ceased. However, we should not overstate the smoothness of this space or the porosity of the borders. To the contrary, ongoing conflicts and harsh obstacles characterize the political relationships between those countries and their migration politics—above all, the Moroccan-Algerian dispute which started in 1994 after terrorist attacks in Marrakech led Morocco to close the frontiers with Algeria and to introduce visa obligations for all Algerian citizens. Despite the period of formal restrictions ending in 2004, today the two countries are far from coming to terms about a liberal politics of mobility: the burning political issue of the annexation of the Western Sahara region to Morocco and the independence of the Sahraouian people that has been going on since 1975 are still at the core of the present quarrels. Moreover, the frontiers are de facto still closed for Moroccan citizens. But the political reverberations of the Arab uprisings expanded also to those countries where the revolution didn't happen in 2011, like Algeria and Morocco. The first step of the Tunisian government was to revise the conditions for the visa and the residence permit for the citizens of Algeria, Morocco and Mauritania. However, despite the multiplication of economic actors beyond Europe their economic agenda is essentially based on many neoliberal measures implemented both by the Gulf States and by the European Neighbourhood Policies renewed in the aftermath of the uprisings and with the entry in the region of economic partners like China. The dependency on and vulnerability to other economies seem to be hampering the construction of an economic alternative in the Maghreb. On 3 March 2014 Tunisia and the European Union signed the Mobility Partnership that was negotiated for more than two years. However, the clause concerning the repatriation of Tunisian citizens is one of the thorniest points that makes Tunisia still reluctant to definitively approve the agreement. In any case, the agreement that also establishes EU support for building a system of asylum in Tunisia also unfolds one of the major spatial transformations of revolutionized Tunisia, which by now has become a space of migration.

Moreover, other transnational spaces of free mobility for trade are supported by the European Union: the 'Arab Mediterranean Free Trade Agreement' signed in Agadir 2004 between Morocco, Tunisia, Jordan and Egypt with the approval of Europe has been promoted by the European Union as an example of the transfer of the European economic integration model to the Arab Mediterranean countries.[24] The ambivalent political issues underpinning the Maghreb space of free mobility compel us to largely rethink both the discourse on free movement and the migration-development nexus in the light of an alternative economy, interrogating what a critique of the developmental paradigm could mean in revolutionized spaces. What is apparent from the context of the Arab uprisings is the inadequacy of a liberal critique that merely advocates for no borders and free mobility: the Arab revolutions have foregrounded and sharpened the necessity to think and act together to critique borders and the productive system. From the standpoint of a long-running analysis, the temporary interruption and short-circuit of the migration governmentality mechanisms that Tunisian migrants enacted has to be articulated within the debate about the construction of a 'new Tunisia'. To sum up, the gaze on migration during the Arab uprisings makes us see the limits of a liberal critique of the border regime, suggesting that a reformulation of critical discourses on migrations from the southern shore of the Mediterranean involves dismantling the colonial and post-colonial imagery which underpins the narratives of democracy.

The political economy of mobility represents an important angle to take for an in-depth analysis of the Arab uprisings. As some scholars suggest (Dabashi 2012; Kanna 2011) migrants' movements and migration politics are crucial elements of the cartography of the Arab revolutions. And at the same time, they have played an important role in workers' struggles that historically have marked the forms of dissidence towards dictatorships in Arab countries in recent years. In turn, as Hamid Dabashi points out, 'the Arab Spring is very much implicated in this tracing of the patterns of labour migrations'. This is the reason why the spaces of free circulation of people that the Arab revolutions can open must be read in the light of 'the proliferation of borders that cut across and exceed existing political spaces' and the related multiplication, read differentiation, of labour regimes as new mechanisms of hierarchization through the 'multiplication of control devices' (Mezzadra and Neilson 2013). In particular, focusing on the *migration roots* of the Arab revolutions, a huge number of migrant workers' struggles took place in the Gulf States in the last ten years, in the form of protest against the exploitative conditions of the labour system. In some way, migrant workers' protests have been the sidelong force of the revolutions. That expression highlights the different position and claims of migrant workers in comparison to the more narrated protests and political unrest of citizens (in Tunisia, Egypt, Barhein, Syria, Yemen) demanding the end of the dictatorships and a real democracy. In fact, as Ahmed Kanna suggests, while migrants did not ask for social or political integration within the perimeter of the polis, citizens made claims and gained their freedom from inside the borders of the space of citizenship. Both these vectors of the revolutions have shaped the ground for social protests, but the struggles of the migrant workers bring to light some limits of citizens' political demands and more broadly of uprisings that still centre on a national framework, as ultimately is the case of the Arab Spring.

The Mediterranean space, which has been shaken by the *twofold spatial upheaval,* does not include only national or land boundaries but also the Mediterranean Sea as a contested space of mobility and as a sea of deaths.

NOTES

1. For a critical analysis of the 'narrative of the revolutions' and European history presented as a blueprint for reading the present events, see F. Sossi, 2013a.

2. The logic of 'learning best practices' which sustains most of the programs of selected mobility grounds on the idea that a smooth transition to good governance requires an 'apprenticeship' of practices of democracies.

3. The category of 'development' is assumed here starting from the analysis of Arturo Escobar, who shows that development discourse emerged in the 1950s as a political project to continue colonial domination in other forms and for taming social unrests in decolonized countries (Escobar 1996).

4. In the village of Zarzis the project for the reintegration of voluntarily returned migrants consisted of supporting fifteen people of the southern regions of Tunisia in starting up economic activities—such as hairdresser, driver or restaurateur—which 'best accord to the profiles and skills of the people '.

5. Cecilia Malmström, 'Responding to the Arab Spring and Rising Populism: The Challenges of Building a European Migration and Asylum Policy', European Union Delegation to the UN, http://www.eu-un.europa.eu/articles/en/article_12133_en.htm.

6. As Sabine Hess explains, 'Countries of transit and origin themselves more and more play the "migration card" in international and economical negotiations. . . . It is getting harder and harder to negotiate readmission agreements with African countries as they start to demand a real equivalent amount for the missing remittances' (Hess 2008).

7. Interview conducted in the city of Zarzis, July 2012, with a young Tunisian citizen.

8. European Commission, 'Communication on Migration', 5 April 2011, http://ec.europa.eu/dgs/home-affairs/news/intro/docs/1_en_act_part1_v11.pdf.

9. IOM was itself put into place in the 1950s precisely to respond to the crisis produced by the two-bloc politics in the aftermath of World War II (Georgi 2010).

10. The expression 'spatial takeover' stands for practices of movements or presences in space that come as unexpected to migration policies and, more broadly, to the established order of visibility. Instead, by '(b)ordering spaces' I mean zones of containment or places for filtering and decelerating mobility.

11. The notion of 'crisis' has gained an economic significance since the middle of the nineteenth century, 'conceptualising emergency conditions, or [situations] related to

class relationships, or produced by the industry or by the market capitalistic economy and that are conceived in their complexity as a symptom of a disease or of an unbalance' (Koselleck 2012, 81).

12. Tunis, 26 March 2013, rejected refugees from Choucha made a protest in front of UNHCR's headquarters.

13. Voice of Choucha, http://voiceofchoucha.wordpress.com/.

14. Interview with UNHCR commissioners at UNHCR office in Zarzis, December 2012.

15. Interview with UNHCR commissioners at UNHCR office in Tunis, August 2013.

16. By that expression I mean the fact that migration policies and administrative measures generate a substantial fragmentation in migrants' mobility that make them 'bounce' back and forth from one country to another.

17. Interview with a UNHCR Officer in UNHCR office of Zarzis, December 2012.

18. Interview with rejected refugees at Choucha camp, August 2013.

19. Interview with refugees at Choucha camp, August 2011.

20. Interviews with rejected refugees at Choucha camp, August 2013.

21. I use here 'subjectivizing' in the Foucaultian twofold meaning of making someone 'subject of' and 'subject to'.

22. Actually, the first negotiations for establishing the Union du Maghreb Arab started in 1964, with the main economic aim of coordinating the development plans of Algeria, Libya, Morocco and Tunisia and the relations with the European Union. However, this plan never came into force until 1989, with the signature of the treaty that officially established the existence of the UMA.

23. In 2010, 54 per cent of labour migrants from the MENA region moved to the Gulf States or to other Arab countries of the MENA region (Paggi 2014).

24. http://ec.europa.eu/europeaid/documents/case-studies/arab-mediterranean_fta_en.pdf.

FIVE

The Desultory Politics of Mobility

Mediterranean Patchy Invisibility and the Humanitarian-Military Border

On 6 September 2012 twelve miles off the coast of Lampedusa and very close to the little island of Lampione, at around 4 PM a boat with 135 Tunisians on board sent out an SOS to the Italian Coast Guard. After more than nine hours 56 of them were saved while the others 'disappeared' in the Mediterranean, despite the tiny dimensions of the island of Lampione. The Italian authorities did not believe the version of the story told by the surviving migrants, suspecting that they had been dumped in the sea by the smuggler, since no sunken ship was found. This sinking of a migrants' boat was not an extraordinary event: since 1988 more than 18,500 people died in the Mediterranean[1]—although the real number cannot be exactly estimated because when shipwrecks are not confirmed by national or international authorities it is almost impossible to count the losses at sea. On the southern shore these uncounted disappearances are much more tangible, since people know who left by boats and never arrived.

Nevertheless, for the first time in Tunisia, in 2011 and in 2012, the families of the disappeared migrants self-organized in groups, setting up protests and political campaigns to find out what happened at sea and demanding that both the Tunisian and the Italian government respond. In some way, the 'mood' of the Tunisian revolution spurred multiple struggles, highlighting the unacceptability of power. These were struggles characterized by a fundamental intractability: 'noisy' practices that could not be easily recaptured by institutional or humanitarian discourses and which staged the refusal to be represented by any human rights association, political party or international organization. Two days after the shipwreck of September in the village of El-Fahs, the families of the twelve migrants who disappeared in the shipwreck of Lampione proposed to the other resident of El-Fahs to declare a general strike; all economic activities were stopped and the arterial road out of the city was blocked. Neither the main trade union (UGTT) nor a political party organized the strike; rather, it was the result of a self-organized network set up by the parents and the relatives of the disappeared migrants. The day after the shipwreck they summoned all citizens for a collective response to the silence of the Tunisian authorities, since neither the national nor the local government communicated any news to the families about the incident, and what was more, they published an incorrect list of the missing people. The general strike represented a gesture of radical distrust towards and a way to delegitimize the government: 'We blocked production so that the government was obliged to see us; and at the same time through that protest we discredited it of any authority'.[2] The words of the parents of the missing migrants addressed the unresponsiveness of the Tunisian government, which did not investigate the circumstances of the shipwreck and which still criminalized 'illegal' emigration. The general strike was the culmination of one and a half year of protests engaged in by the families of the Tunisian disappeared migrants, protests that held migration policies responsible for letting migrants die at sea. This undermined the widely spread humanitarian discourse which denounces the deaths at sea[3] as tragedies that called for a more efficient system of rescue.

The reality of deaths at sea in the Mediterranean is today acknowledged by European agencies and governments, which prompted calls for a better and more coordinated system of security and rescuing in the Mediterranean. While the denunciation of the deaths at the borders of Europe is certainly an important step in bringing out the effects of the 'border spectacle' (De Genova 2013)—building a spectacle of rescue—its capture and translation into the humanitarian discursive frame has somehow tamed the troubling impact of that disobedient gaze. In fact, counter-narratives, critical reports, video and maps which show the 'dark side' of border controls have been, at least in part, incorporated into the human rights discourse promoted by European agencies (Dembour and Kelly 2011). This discourse has tended to shift those counter-maps of the border regime from a critique of the actual mechanisms and rationales of migration governance to denunciations and legal claims against the non-compliance of European States in rescuing people, or for pushing them back. For instance, after the two biggest shipwrecks near the island of Lampedusa on 3 October and 11 October 2013—in which in total 662 migrants died—Human Rights Watch addressed the European Union demanding an improvement of the mechanisms of surveillance and rescue: fostering the link between the two (monitoring and rescuing) is what characterizes the security agenda promoted by nongovernmental organizations, which, in the words of Human Rights Watch, requires that 'an increased surveillance of the Mediterranean, including through the new EUROSUR system, is focused on the paramount duty of rescue at sea'.[4] Therefore, political mobilizations against migrants' deaths at sea were turned by humanitarian actors into claims against the violations of human rights (Feher 2007). Appeals to international law, which obliges government agencies to rescue people distressed at sea, campaigns that claim secure journey conditions for migrants and discourses about the risk of migrating unsafely (read 'illegally') have become the official pillars defining the codification of the strugglefield on migrants' deaths. However, the protests in El-Fahs and the many demonstrations of the families of the lost migrants that took place over 2011 and 2012 both in Tunisia and in Italy cannot be easily captured by the humanitarian script calling for a safer Mediterranean. In fact, through their impossible demand—that the Italian and Tunisian authorities confirm through digital and biometric traces whether those migrants arrived in Italy—they touched the kernel of the two-sided mechanism formed of illegalized movements and selected mobility. Moreover, in the specific context of the Mediterranean crossings the human rights 'equipment', which sustains most of the claims against border controls, has put migrants, I suggest, in a situation of vulnerability as the only (non)choice they have in order not to be stripped of any right to stay is to, effectively, disappear. Being in danger and sending out an SOS call so that they can be detected by the authorities paradoxically increases the possibility for migrants to reach European shores (being rescued and not pushed back).

This chapter focuses on the Mediterranean Sea as one of the Earth's most monitored spaces, and on the ongoing improvement of technologies of control by European states. Against this background, the chapter aims at fraying and flaking the Mediterranean space of mobility, refusing to assume it as a surface or as a place of movement and border enforcements; instead, the Mediterranean is seen here as one of the main metageographical referents of the politics of mobility of the European Union. Secondly, this chapter gestures towards a de-articulation of the safety-and-control paradigm, seeking to undermine the assumptions upon which it is built—the overlap between humanitarian and security concerns—and to disentangle and unpack the all-monitoring logic that underpins the European politics of mobility (Pugh 2001).

The narrative of migration governmentality as an encompassing and well-coordinated regime needs to be deflated and then scrutinized in its specific orientations and mechanisms, as well as in the conflicting instances and interests that cross the supposed horizontal coherence of the international border regime. In order to step back from the image of a coherent system of migration management, I will focus on the effective fragmentation of the European border regime and linger on what I would call the frantic management of the European borders—that is, the tactic of pursuing migrants, which migration agencies are forced to enact in order to monitor their routes to produce statistical reports and to shape risky mobility profiles. The supposed coherent and prefigured set of strategies forming the migration regime, I will show, actually turns out to be a patchy hodgepodge of overlapping technologies. Nevertheless, I don't want to suggest that something like a government of migration does not exist. Nor do I wish to point to the 'failures' of such a governmental regime simply to stress the discrepancies between the discursive regime and the effective functioning of those techniques. Rather, what is at stake is to understand what these 'failures' indicate and stand for: on the one hand, the conflicting interests between different actors—states, private companies, European institutions—and on the other hand, the way in which, despite 'failures', a government of migrations operates. The same point is made about the singular actors involved, such as states that in part oppose the decline of their sovereign prerogatives but at the same time try to take advantage from time to time of some specific measures and standards established by the European Union. The plurality of political and economic actors leads to a scattering and a 'dilution' of political responsibility and, consequently, the migration regime appears to be an autonomous machine without an identifiable agent governing it. Beyond this crucial point, an analysis of migration governmentality at sea should take into account the inadequacy of studying the government of migrations by focusing exclusively on migration policies and techniques of control. The migration strugglefield cannot be fully grasped if we do not overstep the edges of migration as a supposed distinct political and disciplinary field and, instead, locate it at the crossroads of multiple economies of power, including the international labour market, the politics of citizenship, biopolitical powers and the government of conducts (Foucault 2009). What characterizes bordering technologies is precisely the differentiation of status and conditions of mobility of migrants, as well as the ways in which the violence *of* and *at* the borders differently impacts on subjects. Drawing on Foucault, it could be argued that there is no unitary dispositive, or better, that what a critical analysis has to do is just to make visible the non-unity of such a regime, the multiple levels at which it operates and the impossibility of assuming a single logic through which to read the functioning of the government of migration (Foucault 1998).[5] It also cannot be overlooked that if all mechanisms of identification and control ran in a perfectly smooth way, the result would paradoxically be quite counterproductive for states and, especially, for economic actors: the guarantee of a degree of illegality production (De Genova 2002; De Haas 2008) would be paralyzed. From this standpoint, an engaged analysis on migration governmentality should be wary of pointing out or denouncing the 'failures' of governmental mechanisms in order not to fall into the trap of unintentionally fostering the improvement of monitoring systems.

THE MEDITERRANEAN AS A BORDER: THE PRODUCTION OF AN 'UNSAFE' SEA

In order to understand the increasing governmentalization of the Mediterranean Sea it is important to trace back to the moment and the ways in which the Mediterranean Sea started to be moulded as a space of monitored mobility and, at the same time, as an (un)safe space. In fact, the production of Mediterranean (in)security assemblage dates back more than two decades: the shift from the perception of maritime space as a space of rescue and free movement to a zone of interceptions, monitoring and *refoulements* clearly depends on the flourishing of the visa system in the 1990s that strongly limited access to a free and unsanctioned mobility (Cuttitta 2007; De Haas 2007b).[6] 'Since the beginning of the 1990s, under changed geopolitical conditions, the Mediterranean has entered a phase of accelerated juridicalization [. . .] extending national jurisdiction into what used to be high sea' (Heller and Pezzani 2014, p. 665). Thus, the spatial redefinition of the Mediterranean Sea as a border zone, or rather as an assemblage of borders, went along with the redefinition of its legal geography (IMO 1980): Mediterranean controls intensified in parallel with migrants' journeys, setting new legal borders and geographies. In this sense, the Mediterranean Sea is one of the zones in which the dislocation of borders from territorial sovereignty and the current spatial restructuring and multiplication of borders is particularly salient. The high sea was historically considered a space of free mobility, exempt from any form of sovereignty, and this hampered states from intercepting boats in that area. In the last decade, the Mediterranean has become a highly governmentalized space and a contested zone of states' interventions. In fact, the quarrels among states over the areas and the ability of vessels to rescue people or intercept boats on the high seas determine a reconfiguration of sovereignty. As Karakayali and Rigo put it, 'Borders become normative devices that can continuously be reproduced. They do not trace the limits of any given space but reproduce a territorial authority . . . every time that migrants' rights remain anchored to their authorized or unauthorized movements' (Karakayali and Rigo 2010, 138). Both the operative competences and duties of intervention and vessels themselves become mobile borders, even activating the possibility for a state to apprehend suspected migrants' vessels and then relinquish its sovereignty by discharging them to another state: according to the most recent regulation of the European Parliament on the surveillance of the external sea borders in the context of operational cooperation, intercepted boats can be diverted by the ships of a member state towards the territorial sea of a third country, and rescued people should be disembarked to a place of safety—that is, the third country (2013/0106[COD]).

The Mediterranean as a space of governmentality stems from specific technical and juridical measures that trace out discontinuities and produce interrupted migrant geographies making the Mediterranean Sea a border zone. From this perspective, borders are not the demarcating line of political strategies and of spaces of exception, but result from technologies that select, monitor and capture mobility, and work not only by containing and blocking movements but also as regulative technologies (Mezzadra and Neilson 2013; Soguk 2007; Sparke 2006; Walters 2006). The Mediterranean maritime area is at once an anomalous space—because of the indistinct and overlapping border regimes giving rise to controversies between national and international actors—and a spatial lens for enlightening political technologies in which borders and (national) territorial sovereignty are disjointed. In fact, the territorial division between national and international waters overlaps and partially clashes with the operating of mobile borders—such as, for instance, nation-states' patrolling in international waters or in the national waters of another state, as established by many bilateral agreements; through the functioning of radar and satellites; or, finally, tracing zones of blurred sovereignty, for instance, making it hard to understand who is in charge of controlling and rescuing migrants at sea in certain areas (Frontex, nation-states, NATO boats). Non-territorial borders and spatialities have been produced through two main border displacements: first, bilateral agreements and the externalization of frontiers and controls; second, technological monitoring and identification systems.

A quite renowned longstanding controversy concerns the Maltese and the Italian zones of competence in rescuing people at sea in search and rescue areas (SAR). The most famous case, which also involved the Libyan authorities, took place on 29 May 2007, when twenty-seven migrants were found by a Maltese tugboat and rescued at sea. The migrants were not allowed to get on the boat but only to hang on to the tuna cages, where they remained for more than twenty-four hours; after that time the Italian Navy brought them into the detention camp of Lampedusa. Malta had refused to disembark the migrants since the national authorities claimed to have found them in Libyan national waters. A *politics of deferral*[7] is put into place by European states when it is a matter of spending money on rescue operations or taking political responsibility for people left dying at sea. The 'costs of borders' and of border enforcements incurred by states in operating sea patrols and in the 'fight against illegal immigration' is quite indicative of the economy of borders activated around the migratory issue: between 2005 and 2012 Italy spent 1.3 billion euros on border controls, deportations and detentions, of which 283 million came from European funds. However, this number does not include the money paid by Italy to fund Frontex operations at sea, since that cost is kept secret by national authorities. Border conflicts are reanimated among states when they come to dispute the boundaries of their sovereignty at sea: for instance, in May 2011, Malta refused to rescue migrants drowning in its search and rescue area and Italy complained about having to save them; and in August 2011, the Italian government quibbled with Malta over a wider Maltese sovereign search and rescue zone. It seems that the political game of national authorities consists in not being encroached upon by other sovereign states and not acting in their area of competence: they tend to informally discharge tasks and responsibilities to other national and international actors, turning a blind eye to costly operations at sea.

THE TECHNO-POLITICS OF CROSSOVER

It is important to remark that the advanced systems of surveillance were not set into place with the specific purpose of leading a 'war at low intensity' on migrants (Mazzeo, 2011); most of the technologies for controlling coasts and high seas were originally devised for other goals such as controlling fishing and the illicit trafficking of drugs. Many of these systems have been hijacked and used for migration monitoring, or at least, combined with more targeted instruments specifically created for detecting small vessels like migrants' boats. European research studies like Bortec (2008) investigated the feasibility of integrating already existing technologies of surveillance with new systems of surveillance (Kasparek and Wagner 2012; Wolff 2008). For instance, on the island of Lampedusa the biggest radar was built in 1987 to face Gaddafi's attacks. Even the implementation of the radar networks along the coasts is not specifically for the purpose of targeting migrants but it is the result of manifold political and economic goals. The designation of *techno-politics of crossover* well encapsulates the multiple transpositions and political reinvestments of specific technologies of control from one field of government to another. In this regard, the operation of migration controls at sea is far from being a cutting-edge field of governmentality: despite the advanced monitoring tools that in principle should work along the coasts, as well as the supposed smooth coordination between different agencies, the militarized island of Lampedusa reveals another story. The patrolling of the sea is undertaken by each Italian military force acting independently, without any particular coordination or data sharing between the different corps. And although the most advanced radars have a range up to three hundred kilometres, actually those ordinarily used for detecting migrant vessels have a quite limited range of visibility while the detection and rescue of migrants takes place through navies patrolling at sea. However, 'migration crises' are often seen as opportunities for utilizing techniques of surveillance already being used elsewhere: this is the case with drones which, after the deaths of more than six hundred migrants in October 2013 near the Italian coasts, were also used for monitoring the Mediterranean area in order to prevent the tragedies at sea. 'Techno-politics of crossover' also refers to the multiple and combined effects of the militarization of territories that is increasingly evident in the Mediterranean region and that is done in the name of creating a safer space of mobility: monitoring tools for detecting migrants, which have the effect of forcing them to undertake risky journeys to escape controls; the U.S. MUOS radar system based in Sicily—which triggered huge protests for the health risks it poses—and other technologies which ultimately give rise to the insecuritization of space and lives.

In this context, while it is of great salience to shed light on specific forms and contexts of a war on migrants, the paradigm of war, I contend, is not a true grid to capture and explain the effective functioning of migration governmentality (Peraldi 2008). This is not to say that violence does not take place at the borders, or that deaths at sea are mere 'side effects' (collateral damage) of the politics of mobility. Nor does it entail embracing an idea of governmentality as a set of frameworks grounded on non-coercive forms of power and operating through freedom and subjects' agency (Gammeltoft-Hansen 2008). On the contrary, and as explained in the first chapter, this work conceives of governmentality as a strugglefield. The issue here is to question the adequacy of the model of war as the characterizing style and functioning of a governmental rationality that is eminently grounded on controlling and managing migration routes and their temporality—the 'pace of mobility'. The blurring of the functions of many governmental technologies brings together under the generic label of 'fight against trafficking' heterogeneous political objects, among which is immigration. In this work the point is not to reveal migrants' cunning strategies for dodging controls and patrolling at sea, since it would unfolding their strategies of resistance; rather than making migrants' journeys and routes visible, I bring to the fore the array of technical means and knowledges deployed for detecting and counteracting migrants' practices. Nevertheless, it is important not to overstate the discourse about the Mediterranean Sea as an all-monitored and 'transparent' space: indeed, between the supposed overall technological eyes over the Mediterranean and the effective coverage of the Mediterranean area there is a huge discrepancy that does not depend simply on failures in the mechanisms of visibility—especially if we consider the 'disturbances' produced by weather conditions and the so called shadow zones where radars and satellites cannot see. In fact, this uneven visibility in the Mediterranean depends in part on the impossibility of mapping and seeing all movement and presences in space, thus bringing to the fore the limits of representation and mapping (Casas-Cortes, Cobarrubias and Pickles 2011). Undesired migrants' movements are part of the selected politics of mobility that is grounded on the illegalization of all those who are out of the authorized exclusionary channels: thus, the disappearance of some migrants' traces and their undetectable presence is not considered to be of no interest by the authorities.

Actually, if a suspicious boat is detected, it might take more than thirty-five hours to obtain a new useful SAR image of the same boat; or in any case the resolution capacity largely depends on the weather conditions, so that the range for a second possible snapshot could be from twelve to seventy hours later; and given that the average time for a fishery boat to get to the island of Lampedusa from the Tunisian coasts is about twelve hours, it is very likely that the boat would arrive in Italy without being detected again. But the Italian Coast Guard also admits that during the best conditions of visibility the possibility of detecting an 'irregular' small vessel is no more than 80 per cent. Something always escapes the technological gaze deployed by concurrent actors, who, in any case, compete with each other to gain control over a given area of sea. For instance, if in principle real-time information on suspect vessels and on the position of the patrol boats should be shared by all the Italian forces, in reality they frequently quarrel over who needs to undertake the rescue. Taking on the vocabulary of military authorities, once migrants' vessels are detected at sea, in principle they are tracked at a distance along their path—*tracked by shading* is the technical term used by Italian authorities—but this would require a huge deployment of costs and forces, especially when, as in the first half of 2011, the number of migrants' boats per day is considerable.

On the other shore of the Mediterranean, Tunisian patrol boats are in part provided by European states, according to the *do ut des* logic of bilateral agreements, in which the fight against unauthorized movements plays a pivotal role, but they are very scanty in comparison to the advanced technological means in possession of European actors. Thus, the fight over visibility—that is, the striving of migrants to remain undetected and the attempts of multiple actors like Frontex and national corps to detect them—is enacted by governmental forces on the basis of an essential asymmetry between the states involved: North African countries are required to patrol their coasts and to prevent people from migrating, but de facto they act in a condition of enduring dependence on the European forces, since the means provided from the northern shore are not adequate for detecting and then registering into a database all suspect movements. Indeed, fishermen are often questioned to ascertain if a shipwreck occurred, since their constant presence at sea makes them sometimes more aware of what happens. In this way, the support of the European actors remains unavoidable and, at the same time (technological) knowledges are played by European states as a political weapon that is mobilized every time for reiterating the 'conditionality' of the economic and political partnerships (Neighbourhood Policies) and confirming the non-autonomy of the politics of mobility of African states.

The European Surveillance Border System (EUROSUR), promoted in 2008 by the European Commission and officially started in December 2013, is the most prominent attempt to implement a plan of visibility and 'visibilization' of migrants' vessels. This project is presented by the European Union as a sort of coordinator tool that will bring to an end the lack of coordination and information sharing among member states, private actors and international agencies in providing a live map of the movements in the Mediterranean (Jeandesbodz 2011). It is based on a logic of spying and hijacking migrants' routes, in order to trace out a *refractive* and *reactive cartography* of people's movements. Therefore, alleged pre-emptive actions and anticipative risk analyses promoted by agencies like Frontex are actually the result of frantic practices that pursue migrants, trying to hijack, deviate and when possible anticipate their moves. It suffices to cast a glance on how experts and research centres develop new strategies for overcoming the limits and the failures of the 'systems of capture' and detection at sea: 'beyond piracy, also for a wider maritime safety: illegal fishing, immigration, pollution etc.'[8] the Joint European Research Centre declares, and in this way the migratory issue is conflated within a range of other 'criminal activities' which disturb good maritime governance. In other words, it seems that, more than the direct control on specific 'illegal' movements, what is at stake is the enactment of a complex regime of (in)visibility, articulating zones of shadow and subjects of high visibility. At a cursory glance, a fight against opacity underlines all the research and operations aimed at *enhancing the maritime picture*. Thus, what is envisaged by research centres, national authorities and international agencies is a map of visibility that detects migrants' surreptitious movements and spatial strategies, not simply for monitoring but, rather, in order to predict vessel positions and to trace a risk map. However, things are more complex than this, and the image that I suggest of an uneven regime of (in)visibility is useful to explore the ambivalences and the nuances of the 'battle over visibility' that characterizes migration governmentality: the real-time cartography of the movements in the Mediterranean envisaged by international agencies and states ultimately results in a map of opaque patches and spaces out of sight. In fact, as the watchers of the Mediterranean confess, a considerable degree of invisibility is always at play due to variable weather conditions. At the same time, far from aiming at a full-spectrum visibility, the governmental real-time map of the Mediterranean is built on the secrecy of border dis-location. To explain this last point I draw attention to the exclusionary knowledge of the mapping gaze and to the ongoing transformations of the border regime. First, the integrated systems of sea monitoring are based on restricted access, and only the Automatic Identification System is based on an open platform mechanism. Concerning transformations of the border regime, we should look at the dislocation of borders produced by monitoring technologies, as well as by techniques of remote control, biometric data storage (Eurodac, SIS I) and mechanisms of government at a distance like the visa system: as many scholars have long contended (Bigo 2005; Bigo and Guild 2010; Cuttitta 2007) borders are increasingly less reducible to linear edges and tend to become mobile frontiers attached to the body, and multiplying far beyond the geopolitical line (Mezzadra and Neilson 2013). Therefore, on the one hand, zones and intervals of visibility and invisibility are arranged and enacted both by governmental actors and by migrants; and on the other hand, one should underline the technological limits in effectively providing full-spectrum monitoring.

Starting from the multiplication of borders, it should be asked, as Cuttitta does, if one could take on the new borders as vantage points from which to observe and interpret the dynamics of power (Cuttitta 2007). The ongoing dislocation of borders from the national territory should not make us overlook the still considerable impact of state sovereignty in violently blocking people at the borders. But all these analyses are useful in highlighting the way in which the exercising of state sovereignty and of territorial authority have been increasingly detached from one other. For instance, the patrolling of the Tunisian frontiers is made in conjunction between Italian and Tunisian forces. Against this background, the map of the Mediterranean migrations traced by monitoring systems and governmental agencies makes it difficult to understand the dis-location of borders; in fact, what is in place is a technology of tracing and traceability—the traces of migrants' passages, the traces of their bodies—which keeps 'secret' the location of borders and border controls. In this way one could speak of a sneaking power, working underneath the threshold of visibility, which does not indicate in advance where mobile borders are. The fight over knowledge is well illustrated in the secret location of radar stations and the positioning of marine patrols: the map of Guardia di Finanza representing the Italian radar stations is not in the public domain, only researchers and lawyers can demand to see it. And this is just one layer of invisibility, since other integrated systems of patrolling and monitoring remain at the moment untraceable, as with the use of new drones and Frontex operations against immigrants. It is likewise important to underscore the variable geometry of visibility at which power works, alternating politics of detection and non-traceability: the impossibility of attesting to many shipwrecks of migrants' boats and of reconstructing what exactly happened at sea or accounting for the disappearance of thousands of migrants cautions us against assuming migration governmentality as a politics of total visibility and effective operation.

SOFT BORDERS? THE DECEPTIVE HYPE OF TRANSPARENCY AND ACCOUNTABILITY

Since 2011 human rights organizations and European Union agencies have produced a large amount of critical analyses and documents about migrants' deaths in the Mediterranean. Denunciations by activists and human rights advocates of the *refoulements* of migrants' boats in international waters have considerably increased over the last decade: the images of Fortress Europe and of the Mediterranean as a sea of deaths have circulated in the European public debate, at least in its most politicized milieu. However, the impact of all this is not clear. Let's bring the attention to the warnings of the European Council about migrants' vessels and the tragedies at sea: 'Europe's leading human rights watchdog has called for an overhaul of policy on migrants attempting to cross the Mediterranean. . . . In an effort to prevent a similar tragedy from happening again, the Council of Europe has now endorsed a thorough review of existing protocols regarding migrants trying to cross the Mediterranean'.[9] The document, *Lives Lost in the Mediterranean Sea: Who Is Responsible?* has been released by the European Parliamentary Assembly after investigation into the shipwreck of a migrant boat leaving from Libya and the accusation that a NATO boat did not rescue those people in distress: 'There were failures at different levels and many opportunities to save the lives of the people on board the boat were lost. . . . There seemed to be no working agreement between the SAR authorities and NATO headquarters in Naples. This non-communication contributed to the situation in which help was not given to those on board'. This admonition highlights some frictions at stake among European institutions, but at the same time it also reveals to what extent critiques and denunciations of migrants' disappearances at sea are reabsorbed into a governmentality discourse. Therefore, what is at stake is the translation of the critical discourse on migrants' deaths into the institutional domain: the discursive field set by human rights advocates is fundamentally grounded on a firm critique of states and European agencies like Frontex which are responsible for not complying with international standards and for the violations of human rights. Their discourse demands that the conduct of governments be accountable and that operations at seas be more transparent. To better explain the political stakes of the discourse on deaths, I try to articulate the three main overlapping points. First, both institutional documents and human rights advocates claim to be 'filling the void of responsibility': this implies the establishment of a more effective coordination system among the different actors, aiming, in some way, at a smoother functioning of border controls. In other words, it brings out the *double side of visibility*: the claim for a more efficient and legislated system of rescue could easily slip and reverse into the strengthening of the mechanisms of capture.

In this regard, also, the discourse made by human rights groups tends to reproduce and foster the logic of 'by the rule and against opacity'. Actually, disappearances and deaths at sea are neither natural tragedies that governments must prevent nor the dark side of migration governmentality (Grant 2011a, 2011b; Spijkerboer 2007): envisaging them merely as the consequences of the harshest European watchdogs—like Frontex—one overlooks that the 'fight against illegal immigration' is one of the main tenets of the European border regime. Pushing this argument forward, it follows that a critique of the border regime which aims at producing some effective interruptions or disturbances, should move from a focus on migration controls—that many scholars charge with the 'human costs' they implicate—towards a challenge to the visa system and to the partitioning mechanism which differentiates between migrants and asylum seekers. Analyses based on a human rights discourse pushing for a humanitarian border and for softening the 'side-effects' of migration controls (Lutterbeck 2006, 2007) re-inscribe political struggles against the death effect of the visa system into a *watching the watchdogs* logic that points to states' obligations at sea (Lisson and Wienzierl 2007; Tondini 2010). In Tunisia, the critical discourse about deaths at sea and the border regime after the revolution has been centred on a more radical instance: the existence of the visa regime is precisely what needs to be dismantled, since migrants' risky journeys depend on the exclusionary and exclusive conditions established by the European politics of mobility.

At the same time, governmental agencies have replayed the script of security and rescue at sea along two lines: by enforcing technologies of monitoring, as a part of the ongoing militarization of the Mediterranean, in the name of a secure and safe sea; and by partitioning between people taking the risk of leaving in dangerous conditions (economic migrants) and people in need of protection (asylum seekers). These ways of sorting people out, charging some migrants with the responsibility for risking their own lives and presenting others as people to save as asylum seekers, emerged quite blatantly in the Italian context in 2011 with the arrivals of migrants from Tunisia and Libya. During the first months of 2011, when only Tunisians arrived on the Italian coasts, the debate swayed between a humanitarian discourse on people fleeing political turmoil to worry over an unexpected 'wave' of migrants and their unjustified escape. But with the increase of Tunisian migrants on the island and the first arrivals of people from Libya the moral partition became more clear-cut: on the one hand, there were the 'beggars', Tunisians seizing the opportunity of the revolution to escape their country, and, on the other hand, the 'Libyans' claiming asylum. More broadly, the *twofold spatial upheaval* upset and forced a rearrangement both of the order of discourse and governmental technologies of monitoring and capture. In June 2011 the European Council stated that 'these journeys, always undertaken illicitly, mostly on board flagless vessels, putting them at risk of falling into the hands of migrant smuggling and trafficking rings, reflect the desperation of the passengers'.[10] In this way, the crisis of migrants reaching Europe is framed from the outset as the crisis of the European democracies in the face of migrants' upheavals. People leaving North Africa are portrayed as desperate migrants risking their lives in unsafe vessels or through smuggling circuits. And the incorrigible irregularity of migrants is assumed as the condition on which any politics of mobility should hinge, so that 'deaths are nothing but the outcome of irregular movements and unsafe routes' (Grant 2011a). Drawing on the turbulences of historical events—in this case the Arab uprisings—techniques of border controls and migration governance actually seize the opportunity for reassessing or stretching both the rationale of government and the techniques for partitioning and capturing migrants. For instance, in the case of the regime of humanitarian protection: the *migration mess* which took place in the Mediterranean fostered and accelerated transformations in governmental technologies that were partly already in place, like the blurring of the international protection framework and the depoliticization of the logic of asylum. The stress on the category of 'survival migrants' and on the necessity of expanding the borders of protection also to non-refugees emerged with a particular vigour just after the Mediterranean migration turmoil, pushing forward a trend that was already underway. However, the broadening of the space of protection actually leads to a substantial weakening of the asylum, making it harder to get refugee status, which is usually replaced with surrogates like the humanitarian temporary protection.

COUNTERACTING THE MONITORING SYSTEMS: THE PRIMACY OF DISOBEDIENT GAZE IN MIGRATION ACTIVISM

Starting from the patchy spaces of visibility in the Mediterranean, political campaigns and actions against migration controls and deaths at sea tried to come to grips with the visibility regime set up by the mechanisms of migration governmentality. Political or juridical actions, and critical analyses of migrations, are largely predicated on a dissident gaze *at* and *on* the borders: a dissident gaze *on* the borders which assesses the impacts and the tangible effects of borders and shows the side effects of migration management, looking at the border from a different standpoint; a dissident gaze *at* the borders that regards the border as a vantage point for reversing the discourse on migrations and unsettling the order and the thresholds of visibility/invisibility set by migration policies. Thus, this political approach is basically grounded in a demand for more visibility and transparency of power.

In order to expand this argument, I take into account the political campaign Frontexit[11] against Frontex, the European border agency. The campaign against Frontex started in March 2013, demanding the agency respect human rights standards in its operations. 'Disproportionate', 'opaque', 'dangerous' are the watchwords of the video entitled *Europe Is at War against an Imaginary Enemy*, which launched the campaign. The lack of transparency in Frontex's operation and the critique of the border spectacle overlap, finally reinforcing the image and discourse of the migrant as the external enemy that national and European institutions try to chase away. The platform of the associations involved in the campaign demands that Frontex be transparent regarding its operations and that more political and juridical boundaries be set limiting the autonomy of the agency. Advocacy, litigation, awareness and investigation are the four areas of intervention Frontexit engages in, aiming basically at monitoring Frontex activities. Through political activities they demand the agency reject its present guidelines, while the juridical action consists in appealing to the European Court of Justice and the European Court of Human Rights, 'using the legal avenues to bring to light Frontex's responsibilities as regards the violation of migrants' fundamental rights'. In this regard, it is to be stressed that while the first campaigns against Frontex, like '*Frontex explode*', targeted the existence of Frontex as such, promoting and undertaking actions to sabotage its activities, Frontexit now focuses rather on the necessity of limiting and opposing the autonomy of Frontex's actions, demanding transparency from the agency. The principle of the 'democratization of the borders' (Balibar 2004) seems to qualify the campaign, translating the protests against the very existence of the agency into a redefinition of its boundaries of visibility, pushing for a democracy of the visible. If, on the one hand, the multiplicity of European institutions allows campaigning against an agency created by the European Union itself, on the other hand, it remains the fact that Frontex is not an exceptional entity generated by Europe but an agency conceived for implementing cooperation on border controls among member states. What characterizes Frontex is precisely the blurred line between the relative autonomy of its conduct and legal personality and its status as a European cooperation agency. And as I will show in the next chapter, this political and juridical ambiguity leads to the factual impossibility of determining the legal responsibility between member states and Frontex about human rights violations and deficits in the implementation of the duties of rescuing people at sea.

ILLIBERAL PRACTICES WITHIN THE LIBERAL LAW AND THE ORDINARINESS OF OUT-OF-LAW PROCEDURES

The call for transparency needs to be framed in a broader political and juridical rationale: the politics of visibility for fighting human rights violations at sea hinges on a supposed sharp division between the rule of law, on the one hand, and 'illiberal practices' or arbitrary powers on the other. In this way, the 'watching the watchdogs' gaze is basically grounded on a liberal political horizon that regards the boundaries of the law as the guarantor of power's legitimacy; and consequently it presupposes that liberal law can re-instantiate a space of fair governmentality. Indeed, analyses of exceptional zones and states of emergency tend to obscure recognition of the fact that the functioning of migration controls responds to uneven legal and administrative regimes in which illiberal practices sustain the existence of ordinary laws. A gaze focusing on exceptional measures, and attentive to the abuses of sovereign power, tends to overlook the 'banality' of border controls and the ordinary low-intensity measures through which migrants are identified, classified and subjected to 'illiberal' techniques of detention and deportation.

'MAPPING OTHERWISE' AND MAKING 'ANOTHER MAP'

Against this background I would like to address these forms of disobedient gaze, questioning the possibility of producing interruptions in the mechanisms of power (Pickles 2004).To what extent do they effectively disrupt the cartography of calculated bodies/practices? Or, rather, do they reassess the limits of power and 'map otherwise' migration governmentality? Drawing on the idea of politics as the breaking of the ordinary and expected location of bodies in space, the question is whether or not a practice of 'mapping otherwise' finally works within the borders and within the game of visibility set by migration governance. The strategy of demanding that states and international organizations account for their political responsibilities is crucial for reversing the position of being governed (Chatterjee 2004; Foucault 1997): demanding that advanced technologies that monitor mobility make their images available to us means trying to reverse the logic of securitization that is commonly presented as a guarantee for securing our lives. But this gesture tends to take for granted pre-established political frames, addressing the European institutions as guarantors of respect for human rights standards. In fact, a 'counter' approach coupled with a battle over visibility implies that these political entities are recognized as stretching the thresholds of the tolerability of power and that national authorities are to be denounced for their 'failure' to fairly govern migrants. Meanwhile, European agencies stress the need for a real-time situational picture of the Mediterranean area—with the aim of more efficient border surveillance and intervention in rescue operations. This focus on the visibility strugglefield at sea raises a broader question: what could it mean to trace a counter-map once governmental actors also promote real-time updated maps? The idea to build an informal network of assistance proposed by some platforms of activists should take account of the ambivalent strategy promoted by some European agencies. According to what I called a *politics of discharge*, European expertise recommended that fishermen and private shipmasters should not be sanctioned for providing assistance to migrants. In this way, a politics of deferral along with a discourse of civic responsibility operate at the same time: on the one hand, institutions, most of all in a time of economic crisis, try to minimize the costs of border controls, while on the other, they reverse the social and moral responsibility on civil society of the tragedies at sea. However, the struggle against opacity that governmental agencies promote in the name of security and safety at sea goes hand in hand with the concealment to the public of push-back operation on the high sea.

Therefore, the argument that I advance here is that the humanitarian discourse puts at the core the reality of the deaths at sea, establishing a conceptual link between the loss of lives, the danger to migrants and the risky choice to migrate clandestinely. Techniques of surveillance are postulated as deterrents against the circuit of 'illegal' migration; in this way, the humanitarian domain spans from rescue operations to the fight against the smuggling economy, since the act of saving migrants' lives is posited as a way to subtract would-be migrants from dangerous circuits.

THE VIOLENCE OF/AT THE BORDERS

Both the production of counter-maps marking the 'dark side' of border controls and the discourses which counteract the narratives that depict migration as a threat for nation-states centre around the twofold paradigm of violent borders and violence at the borders. This entails that border-zones are regarded as exceptional sites of blurred sovereignty, where the political responsibilities of governmental agents become confused and the standards of human rights and international laws are very often infringed. We can concur with Foucault in saying that the notion of violence is not useful for describing the exercise of power since, by centring on an 'excessive' or unregulated force, it leads to the faulty conclusion that not all powers are physical and apply to bodies, and also that rational and violent powers exclude each other (Foucault 2006a). Moreover, we might add that such an image of 'excessive' power ultimately suggests that there could be something like a legitimate exercise of power, while, as Foucault indicates, any form of power can become intolerable and thus object of a refusal.

Moreover, the respect of norms and international standards is used as the yardstick to evaluate whether migration governmentality plays fairly and by the rules. Focusing on the 'supplement' of violence exercised at the borders, and denouncing the violation of the laws that governmental institutions are expected to abide by, the constitutive sabotage exercised against undocumented migration eclipses at the advantage of a 'convertible violence' (Balibar 2010a): in other words, an exclusive focus on the 'excessive' violence—human rights violations—overshadows the violence that is constitutive of the very act of tracing borders and instantiating differential rights to mobility. That kind of violence can be easily converted into accepted technologies of bordering, political statements and humanitarian-securitarian measures. In this regard, Balibar's reflections in *Violence et Civilité* can be explored in order to disentangle the question at stake here, bringing attention to what Balibar calls the transformations of violence. According to Balibar what characterizes present societies is a form of violence that cannot be translated into political codes and that, by its nature, exceeds all antagonisms and conflicts, working rather as a permanent horizon and a condition of their deployments. In this sense, political philosophy obscures what exceeds the boundaries of the convertible violence, revealing its irreducible character. This unconvertible violence, Balibar contends, assumes today two complementary forms: ultra-objective violence consisting in mechanisms of de-individualization, reducing human beings to things, and ultra-subjective violence, which 'represents some individuals and groups as the personification of the evil' (Balibar 2010a, 86). Coming back to migration governmentality, the violence *of* and *at* the borders is to some degree comparable to the unconvertible dimension posited by Balibar: far from being only a transgression of norms and laws to be re-codified, the constitutive sabotage of migrants' practices represents the frame and the horizon of migration management. In other words, in the government of migration something resists any codification in terms of violations of norms and rights: the turbulence of migration reveals precisely the untenability and illegitimacy of the cartography of temporal and spatial borders against which migrants are supposed to move. It is not in terms of 'failures' that migration policies should be challenged: the failures in keeping up with migrants' rights, the failures in rescuing people in distress at sea, the failures in setting up fair mechanisms of government and so on. All these 'failures' are not voids to fill up; rather, they need to be read in the light of the patchy visibility of the migration regime, in part because of the technical limits of monitoring technology and in part because the fact of not-seeing is definitively one of the ways for selecting migrants (Heller and Pezzani 2014).

However, the constitutive sabotage and violence enacted by governmental forces doesn't work solely through exceptional, disproportionate and illegal exercises of power, albeit these aspects are certainly at play, both on the scene of the border spectacle—as in Lampedusa—and offstage, letting migrants disappear in the Mediterranean (Sciurba 2009; Sossi 2007, 2012). The unconvertible dimension of sabotage and violence of migration governmentality consists, rather, in the very principle of a differentiated and selected politics of mobility, which partitions people's movements into degrees of (il)legitimate mobility and profiles of (un)accepted subjectivities. It follows that a critique of the violence *of* and *at* the borders should be situated less within the broad paradigm of a war on migrants than in the ambivalent and exclusionary mechanisms of the politics of selected migration. At the same time, the challenge must not be simply to produce a detailed map, in which the failures and the violence of power are made visible. To put it differently, a counter-mapping approach should not fall into a circular critique of violence, assessing the legality or the illegality of power's techniques or the proportionate correspondence means-ends (Benjamin 1986). As Judith Revel rightly stresses, the real issue in talking about violence 'is to tease out how violence plays within a specular configuration of power'. Furthermore, 'the mutual inducement of powers and resistances . . . doesn't mean that they are of the same nature. Rather, the challenge consists in breaking the circular movement between powers and counter-powers that the matrix of violence instantiates' (Revel 2013b).

'Another map' does not mean only to shift the gaze to unheard or invisible practices; it also means refusing to assume in advance the meaning of struggle itself instead of testing it through migrants' actual strategies (Mezzadra 2013). If the practice of 'mapping otherwise' makes us see the 'other side' of border management, the production of 'another map' dismisses from the very outset the legitimacy of borders and refuses to be in step with the bordering of spaces as well as with the existing political space of address.

The deterritorialization of borders and the increasing differentiation of their functions and topology should not be considered in itself in terms of a smoother circulation and less constraints to freedom of movement (Karakayali and Rigo 2010). To the contrary, the point is precisely to disconnect bordering processes from the geopolitical matrix: beyond the operation of power at the borders, violence *of* and *at* the borders also consists in the temporal management of migrations—the production of an anticipatory cartography of movements and the efforts to spatially trace, register or disturb some specific forms of circulation and presence in space, such as 'irregular' migration. Nevertheless, an analysis of the transformations of the (b)ordering practices should not disregard forms of violence perpetuated in certain spaces or upon specific mobility profiles. Such violence could take the form of de-individualization processes, producing a surplus of unproductive monitored (im)mobility that is exercised by leaving some subjects 'off the map' (Neocleous 2003)—for instance, letting them die at sea.

THE CONTESTED 'RIGHT TO LIFE' AND THE BROADENING OF THE NOTION OF SURVEILLANCE

The humanitarian-security script blurs the constitutive violence of the mechanism of selected mobility which makes some disappear in the Mediterranean while allowing others to leave safely, stressing instead more evident forms of violence. It also focuses attention on the violation of international norms that human rights advocates are prompt to condemn. The current debate on deaths at sea constantly refers to the right to life enshrined in the Charter of Fundamental Rights of the European Union. The right to life of migrants as human beings (before becoming migrants) is presented as a universal right, while the condition of being migrant is implicitly envisioned as a fault that doubles and reconfigures the human condition. In this regard, the first thing to notice is that the floating meaning of 'right to life' embraces and blurs at the same time the multiple occurrences, forms and meanings of 'life' shaped and governed by migration policies. To put it differently, the general notion of life, constantly mobilized for invoking humanitarian obligations, tends to cover an array of meanings and forms of life that every subject is expected to embody (Douzinas 2007). In particular, the floating meaning of life as used in the specific context of migration policies works as a sort of redux factor, assuming life as a 'life on the edges': migrants arriving by boat are constructed as risky subjects, 'risky' referring both to the hazard through which they choose to jeopardize their own life and to their condition of life in the country of origin that facilitates their being trapped in the circuits of smuggling. As Ruben Andersson points out, 'Sea surveillance depends upon a framing of boat migration as dangerous by definition, a risk to life' (Andersson 2012). But more than that, migration policies shape forms of subjectivity and involve specific meanings of life: they instantiate selective and filtering borders by enacting sorting mechanisms—partitioning between profiles of mobility—that produce forms of differentiated subjectivity, corresponding to different meanings of life that subjects are entitled to live. Ultimately, we concur with Judith Butler contending that 'the frames that work to differentiate the lives that we can apprehend from those we cannot not only organize visual experience but also generate specific ontologies of the subject' (Butler 2009, 3). We can similarly suggest that migration policies differentiate among forms of life that different migrant subjects are expected to live. However, under close scrutiny what seems to characterize the discourse and the measures of the humanitarian-securitized politics of mobility (Balibar 2013) is not so much the lack of recognition of the precariousness of migrants' lives as a use of it as a tenet for requalifying migrants' life as a mere question of survival, and thus conflating the right to life into the right not to be left to die. The production of differential borders and rights to mobility implicates by design a differentiation among conditions, levels and forms of life that migrants, non-migrants, refugees and so on are said to have right to. After all, the politics of 'humanitarian-securitization' of the Mediterranean Sea stresses that the only indisputable right that migrants are entitled to, irrespective of their juridical status, is the right not to be left dying by the same policing mechanisms that block and contrast their movements. The empty signifier of 'life' obscures the different regimes of life in which migrants are situated. Beyond the violence of and at the most visible borders, a notable example consists in borders through which would-be migrants are fixed in space, in the country of origin, both by the visa regime, by unemployment or by 'legal geographies as powerful instruments in locating people in space' (Basaran 2011). Thus, migration politics contributes to sorting among different forms of liveable life, which are then translated into different degrees of legitimate mobility. The general invocation of a right to life for migrants in distress at sea reorients the gaze towards seeing migrants as subjects rescued from natural accidents and from their own risky choices (Balibar 2013; Pugh 2004). As William Walters has nicely put it, the life that becomes an object of the humanitarian-securitarian government of migrations cannot be addressed in terms of pastoral power as described by Foucault; the floating signifier of 'right to life' is mobilized neither according to a rationale of care nor on the basis of a logic of individualization (Walters 2011a).[12] The humanitarian discourse can be truly understood only by taking into consideration the spreading of bordering techniques in the name of more security and safety together with the politics of dis-charge that intervenes selectively and discontinuously.

THE DESULTORY POLITICS OF MOBILITY AND THE MILITARY-HUMANITARIAN BOND: MARE NOSTRUM BEYOND THE SEA

To better understand the way in which the human rights discourse has been repositioned through the politics of control in the Mediterranean and which meaning of human life is actually postulated in the logic of securitization, I start also from a specific event. On 3 October 2013, 366 migrants coming from Libya died in the waters close to the Italian island of Lampedusa; 155 were rescued. On 11 October 2013 another big shipwreck occurred between Malta and Lampedusa; 268 people died. In both cases the Italian authorities were accused of fatal delays in rescue operations. The Italian government declared a day of national mourning for the tragedies at sea, without mentioning the mobility restrictions of the visa regime that force people to take a boat and cross the Mediterranean. Just one week after the second shipwreck, the Italian Home Office and the Ministry of Defence launched Mare Nostrum, 'a military-humanitarian operation' in the Mediterranean for rescuing migrants at sea. It is noteworthy that the mission has not been designated as a securitarian operation but rather as military-humanitarian, in which the first term entirely redefines the second. In fact, only in the following lines does the document report that 'the aim is to increase the level of human life security and the control of migration flows'.[13] The concept of security essentially remains in the background of Mare Nostrum's actions and is not helpful for understanding the transformation of the Mediterranean Sea into a space of patchy governmentality—that is, as a space in which zones at high density of border control alternate with others in which bodies may pass undetected. In particular, there are three relevant slippages concerning the meaning of security and its articulation with human rights that, I contend, characterize the production of the Mediterranean as an (un)safe space of mobility. However, this can emerge only insofar as we step out of texts and documents that present Mare Nostrum as a military-humanitarian mission: only by investigating closely the effective actions of the military forces and hearing the stories of the migrants who survived the shipwrecks is it possible to fully unpack the script of security and look at the political technology which effectively is at play.

First, although the traditional governmental field of security cannot be detached from the production of insecurity and from a sense of 'unease' (Bigo 2005, 2008)—as is the case when 'illegal' migrations become part of the 'border spectacle'—in the current 'military-humanitarian' operations at sea this production of insecurity is far from the primary outcome. Indeed, the goal of what I would call the *military channels of rescue* is rather to subtract the island of Lampedusa from the spectacle of migrants' arrivals, rescuing migrants close to the Libyan coasts and disembarking them in Sicily. The continuum of threats formed by terrorism-migration-criminality-trafficking does not effectively illustrate the rationale and the functioning of the patchy migration governmentality in the Mediterranean: especially after the outbreak of the Syrian conflict and the consequent pre-eminence of a discourse about activating humanitarian channels for Syrians, migrants from Libya are accounted in the 'military-humanitarian' rationale less as subjects of risk, but as subjects *at* risk. They are depicted at the same time as subjects who are at fault by putting themselves in danger—leaving with unsafe boats—and as subjects who need to be saved. In this context, the human rights discourse in some way *fades into the sea*: while the humanitarian logic usually relies on human rights standards for opposing third countries' political governments—denouncing, for instance, the conditions in Libyan detention centres—when it is transposed into securing migrants' lives at sea it is immediately reframed as an affair of military concern due to its exceptional character; and the very notion of 'human' is translated into 'life to be rescued'. Second, the rights of migrants' whose lives are at risk are actually rights that reflect on the states' duties not to let them die or be put in danger. Indeed, the obligation for states is to not turn back migrants on high seas, and to afford them the right to protection against returning to a country where an individual faces the risk of torture or persecution.

Therefore, human rights are removed from the possibility of serving as a strategic foothold for migrants' agency, since they come to designate dangerous conditions that people must be protected from (such as the right of not being turned back) instead of designating specific freedoms to be granted. In this sense, the military-humanitarian politics of saving lives at sea contributes to translating human rights into the humanitarian rights of those subjects who must not be left to die (Brown 2005; Rancière 2004). Human rights at sea function like the impassable limits of governmental action—for instance, the duty of *non-refoulement*—and become yardsticks for people's spatial relocation: in the name of respect for human rights migrants can be disembarked in certain states, or they must be allowed to arrive in a 'safe country'. In this regard, it is worth recalling Žižek's reflection on the paradoxical and exclusionary character of human rights that emerges precisely when we are confronted with people deprived of any rights except to be humans: 'Paradoxically, I am deprived of human rights at the very moment at which I am reduced to a human being "in general", and thus become the ideal bearer of those "universal human rights". . . . Far from being pre-political, "universal human rights" designate the precise space of politicization proper' (Žižek 2005, 131). Thus, the rights of migrants' whose lives are at risk are actually rights that reflect on states' duties not to let them die or be put in danger.

In fact, the securitarian military apparatus (monitoring tools and techniques of surveillance) becomes a sort of safety operator, aimed at granting the integrity of migrants' bodies. As I stated above, it is only out of the text that it is possible to touch upon the effective mechanisms of capture of migrants. On 21 March 2014 the *Sirio*, a ship of the Italian Navy docked at the harbour of Augusta in Sicily after two nights on the high sea, where it rescued 340 migrants who left Libya by boat. 'Number 68, look here, on your left'. A policeman took a photo of the migrant's face and then asked him name, age and nationality. The first identification process stopped here; fingerprints were taken later, in the hosting centres. The 340 migrants were split into different groups according to the order in which they got off the *Sirio*. Those who refused to be identified were taken aside and not allowed to leave the dock. 'Where are we, are we in Rome or in Lampedusa now? And where do they take us?' Nobody answered one of the Eritreans, who were all still waiting on the deck. Nor were asylum processes explained to the migrants, who undertook the pre-screening with the police. The 'good' border spectacle that saves migrants' lives at sea stops just at the dock. And from there begins the 'ordinary management of unplanned migration flows', to quote a national directive.[14]

The team of the Italian Navy based in Lampedusa is aware of its new task: 'We are in charge of humanitarian operations now; after all, people choose to live in extremely dangerous conditions and consequently their life cannot be saved other than by highly equipped forces as only the army is. We operate in some way as a ferry-boat for migrants'.[15] The *military migration ferry-boat* is ultimately part of what I would name the *exceptionality of (migrant) mobility*: once again, more than presenting it in the frame of security or of a threat continuum, unauthorized migrations are envisioned as mobility to the limit, namely, people moving in exceptionally dangerous conditions and, as a consequence, requiring the mobilization of military forces. Migrants' mobility is exceptionalized, as it cannot take place without fatality other than through the deployment of an 'economy of safety' (mobilization of military forces, huge economic expenses[16]). This economy of power that exceptionalizes migrants' mobility cannot but bestow on the military the humanitarian task itself ('who, except for us, could save lives on the high sea?') reshaping in this way both the military and the humanitarian at the same time: the humanitarian at sea leaves the subject of human rights in the background and conflates them with the saving of lives; while militaries are, in turn, reconfigured as now being in charge of avoiding migrant deaths. Nevertheless, the politics of monitoring-and-saving migrants at sea cannot be taken per se without looking at the stages which follow the scene of the military-ferry: in fact, once rescued migrants are identified and fingerprinted on the boats, and then, soon after their disembarkation in Sicily, they are taken into hosting centres—where many of them wait for months before their asylum claim is processed—and some are also directly deported. Therefore, by deploying the military-humanitarian operation in its temporal continuity and following it beyond the sea space, it becomes evident that it is not a question of building smoothly managed channels of mobility but of taking migrants into the territory and there commencing the usual exclusionary and arbitrary relocation of people.

However, if one looks beyond the official humanitarian discourse, one realizes that almost nothing has changed at the level of techniques of surveillance and monitoring with the commencement of Mare Nostrum: the radars and methods of patrolling are the same as before Mare Nostrum. Rather, the 'military-humanitarian' logic has recast already existing technologies and operative tools by displacing the issue of human rights to centre instead on lives to save—more than to securitize.

> Before 3 PM we sent an SOS call to the Italian authorities and we had been told that we were too far from Italian water and that we had to go closer to them. . . . Around 4 PM we saw a plane monitoring the zone and thus we began signalling our presence; only after 5 PM did we start to be rescued. . . . We stayed three days on the military boat before disembarking in Sicily. . . . There, once in the detention centres, most of us were forced by violence by the police to give our fingerprints, trick-

> ing us saying that then we would be free to go. . . . After five days Italian authorities feign not to see us and let us escape the centres, and all of us took the trains towards Milan, in order to arrive in Germany".[17]

S. is one of the survivors of the shipwreck of 11 October, and his reconstruction of the events sheds light not only on the delay in rescuing operations but also on what happened after disembarking in Italy—in particular the arbitrary administrative measures that neither responded to a homogenous rationale nor actualized what governmental texts dispose.[18] Most of all, this story allows us to question the image of an all-embracing politics over (migrants') lives. Indeed, as illustrated above in the chapter, the frequent shipwrecks occurred despite the high level of 'securitization', and the failures in assisting migrants in distress indicate the patchy governmentality that characterizes the humanitarian-securitarian gaze and the politics of dis-charge in relation to migrants' lives (migrants are at times object of advanced controls and at times are left dying at sea). Therefore, is biopolitics the right name to designate mechanisms of governing conducts that work primarily by taking hold of life but do not actually capitalize on life? In fact, although 'life' is at the core of the humanitarian-securitarian discourse, and all interventions are made in the name of saving migrant's lives, the meaning of 'life' that is actually postulated corresponds to migrants' bodies to be rescued and, jointly, 'life' that puts itself at risk. Moreover, migrants are subjected to a desultory and irregular monitoring eye, which neither values life nor effectively manages the circulation of movements. On the contrary, migrants' movements are constantly fragmented and blocked. The *desultory politics of (im)mobility* in the Mediterranean is accompanied by a patchy governmentality that sees, monitors and rescues at times (with many grey zones of invisibility), and which operates through a coexistent double move: blocking-rejecting, on the one hand, and channelling people through rescues on the other. In this sense, as Michael Dillon rightly puts, it is necessary to interrogate the slippages of the meaning and functioning of biopolitics today, scrutinizing how the very referent of 'life' has changed over the last decades and whether it is still appropriate to designate biopolitics as a technology of protection (Dillon 2007; see also Selmeczi 2009). Nevertheless, it is important to notice that in the context of the desultory migration politics of (im)mobility what is at stake is not a mechanism of life empowerment but, rather, the maintaining of a fragile line between not seeing—that is, *letting people die*—and monitoring for channelling and blocking people—*making people not die*.

The second important element of confrontation to consider is that if we take Foucault's account of biopolitical power, and also the recent analyses of its transformations, the co-implication between dispositives of security and freedom is presented as one of its most seminal features.

Now, the military-humanitarian operations to 'save migrants lives' actualize in a technology of government over migrants' bodies in which freedom is ultimately *offside* and evacuated: in fact, freedom is not the correlate outcome of more security—namely, more restrictions and controls on migrants' movements—nor does it work as a necessary element for the functioning of securitarian mechanisms. Quite the contrary, freedom remains the uncontemplated possibility for rescued-secured migrants. Indeed, in the humanitarian-military discourse freedom is absent from the start: forced to leave their country, the rescued migrants put their life at risk crossing the Mediterranean by boat, and then, once rescued, they cannot but follow the established humanitarian channels, being allocated to a certain space in such a way that no future movement and life project could be planned. The others, those becoming economic migrants, are de facto criminalized and deprived of liberty for having enacted their freedom of movement irrespective of states' regime of selected mobility.

Eventually, the *biopolitics of the governed* might be a term for addressing practices of refusal and strategies of resistance that try to evade the capture of lives and the desultory mechanisms of control; indeed, these are not enacted on a specific (national) population but crosscut different spaces and subjectivities, producing many sites of governmentality. Different and provisional populations of the 'governed' seem to be the hazy object of the politics of control over migrants' lives, which does not shape and target a bounded population but, rather, gives rise to an array of people subjected to the same mechanisms of capture. Actually, it is not unusual for migrants to refuse to be identified: no photos, no fingerprints, thanks; I want to move away, somewhere in northern Europe. Their refusal is an expression of the desire to wander across Europe. As one of the migrants who arrived in northern Italy stated, 'For the moment I stay here, but then I would really like to wander, to go around a little bit, France, London, Germany and then maybe come back. What do you think? Is it not good to travel?'[19] In fact, despite their forced and peripatetic journeys to escape war conflicts, the subjective desires once they arrive in Europe cannot be fully 'channelled' and managed through 'humanitarian-military' convoys, and it is precisely what puts the migration sorting machine in difficulty: beyond the sea and the rescue, people try to disengage from the humanitarian grasp that strands people in spaces and channels migrants' lives into patterns of (un)protection. Thus, far from being at hand, the spaces of freedom that the rescued migrants try to open emerge from the strugglefield in which migrants are situated. Thus, spaces of freedom are the provisional outcome of a 'strain' that sometimes migrants succeed in producing against the disciplining of bodies and movements, by dodging or troubling the meshes of governmentality and loosening its hold—*border interruptions* that unsettle the humanitarian machine with bodies, movements and subjective drives that cannot be fully regulated and that highlight the frantic attempt of bordering captures—migration policies, states' restrictions—to encode 'disordered mobility'. On 20 March 2014, 250 Syrians and Palestinians escaped the tent-camp in Porto Empedocle, one of the Sicilian harbours used by Mare Nostrum, and spread across the country. Like many others who escaped and then wandered in the Italian towns, they refused to be identified in order to escape the territorial trap imposed by the Dublin III regulation.

February 2014, Bologna: 'We disembarked in Sicily, after two nights, and then we were moved very quickly to Catania, by bus, and then to Bologna, by plane. At the airport we were partitioned: some, like me, in Bologna, another group were taken I don't know where'.[20] While Mare Nostrum rescue operations are reported on the official website of the Italian Navy, nothing is known about what effectively happens after migrants are disembarked on Italian territory. In some way, *the space of visibility is bordered to the sea*—that is, the military-humanitarian operation is presented as ending as soon as people disembark. Where migrants are taken onto the mainland, how they are identified and how long they wait before receiving an answer about their asylum claim is something that remains unclear. Officially people are taken to Mineo, the biggest hosting centre in Italy, or in informal processing centres in Sicily. However, in the city of Bologna, one thousand kilometres away from there, in February 2014 rumours started to circulate about the presence of eighty Malians coming directly from the harbour of Augusta. And after meeting a few of them in the city centre during their daily exit from the hosting centre their location became clear: they were hosted in a building on the hills in the premises of the city, but their presence remained hidden. 'We were fingerprinted on San Marco boat, then it took two days before disembarking in Sicily; and then we were fingerprinted again here in Bologna'.[21] After the official ending of the North African emergency declared by the Italian government in February 2011, and with overcrowding of the hosting centres, migrants were spread over the territory and put in the so called hub centres; their presence in the Italian cities is known only when people start to see them around, passing unnoticed by the most of the hurrying pedestrians. 'We are parked here, I have no idea for how long; there is nothing to do, no job, anything. But I will wait for a little bit here, to restore myself, and then I will cross the border to go to Germany, I think'.[22] Migrants were stranded in different towns or in isolated places for an indefinite time, despite the short-term hosting established by the Italian government. While the military-humanitarian mission was presented as a planned strategy to manage people at sea, on land, the presence of the same migrants has unexpected results.

This insight into the *desultory politics* over migrants' lives casts light on the differences between the humanitarian government in the (refugee) camp and at sea. Indeed, a focus on humanitarian government at sea allows for an unsettling of the fixed and bounded space of the camp. Instead there is a concern with mechanisms of migration governmentality that produce and are grounded on spaces on the move, namely, temporary spaces that variably function as spaces of protection or containment. In fact, in order to understand the functioning of the desultory politics over migrants' mobility it is necessary to draw attention to temporary spaces of governmentality—where people are rescued, channelled through or left to die—also formed by moving transports, like military navies. Moreover, it requires mobilizing a dislocating gaze that takes migrants' displacements as a vantage point to come to grips with the biopolitical holds and transformations to which migrants are subjected. Indeed, it is only by following migrants from the southern shore of the Mediterranean to their arrival on European territory that it becomes possible to see that something like a homogenous and continuous humanitarian government does not exist. Instead of pointing to an overall humanitarian rationality that manages migration 'from one shore to the other' of the Mediterranean, it is far more salient to look at heterogeneous techniques of dis-charge that take a varying hold on migrants' lives and shape the humanitarian profiles that differ from one border to another. For instance, the same migrants 'change' as far as their status and their subjective interpellation are concerned—on the boat they are rescued lives, upon arrival in the territory they become asylum seekers or migrants to deport and at sea they are addressed as subjects at risk. Moreover, the humanitarian grasp works differently—through heterogeneous techniques and temporalities of intervention—according to the different bordering spaces (i.e., the boat, the moment of arrival, the identification procedure). Therefore, it could be suggested that focusing on migration governmentality at sea entails following the 'migration of the humanitarian' in its different stages and across spaces—in a nutshell, well beyond the sea. In fact, a gaze upon Mare Nostrum and on the discontinuous holds on migrants' lives, as well as the different mechanisms of capture and management, highlights that the desultory politics over mobility at sea largely oversteps the boundaries of the sea.

Indeed, as this chapter shows, the patchy visibility of migration governmentality, the peculiar notion of 'life' that underpins biopolitics at sea—rescuing bodies—and the object of control—tracking movements and routes—is quite distinctive of the military-humanitarian politics of managing migration in the Mediterranean Sea. On the contrary, in the context of camps like Choucha the politics of number is constantly intermeshed with the collection of personal stories, and people's mobility is monitored through steadier eyes. The exceptional-risky status migrants' mobility produced by the exclusionary mechanism of the visa system and by migration policies at large is finally actualized in the 12 million euro cost per month of Mare Nostrum operations. However, this relevant data should be used with circumspection: indeed, the most popular criticism against the military-humanitarian operations focuses on the fact that 'this money is used to save migrants in the place of Italians', and 'in this way more migrants arrive on the coasts since they are aware that they will be rescued'. Confronted with that, we cannot but counter those criticisms by contrasting migrants' lives to the lives of others, prompting the possibility for everybody to freely move and leave a country without being at the mercy of a desultory humanitarian-military gaze.

NOTES

1. United Against Racism, 'List of Deaths', www.unitedagainstracism.org/pdfs/listofdeaths.pdf
2. Interview with the one of the parents of the missing migrants, El Fahs, December 2012.
3. It is important to stress that in parallel to these analyses focused on the Mediterranean 'border zones', critical studies of migrants' deaths at the borders are growing concerning other areas as well. In particular, the U.S.-Mexico frontier and deaths at sea in Australia (see for instance, Cornelius 2001 and Stephen 2008).
4. Human Rights Watch, EU: Improve Migrant Rescue, Offer Refuge, www.hrw.org/news/2013/10/23/eu-improve-migrant-rescue-offer-refuge
5. On this point see in *The Will to Knowledge* the chapter on the dispositive of sexuality, in particular the section on the unity of the dispositive where Foucault remarks this point very clearly.
6. As Hein De Haas stresses, the first visa restrictions concerning North African countries were introduced by those countries in the 1960s and 1970s, to prevent people migrating to Europe. In fact, the crime of 'illegal emigration' that Tunisia introduced in 1975 and that is still now in place was a measure for controlling and reducing the huge outflow of Tunisians citizens towards France, since with the oil economic crisis of 1973 and the consequent crisis of the European labour market, Tunisia turned its economic interests to countries like Libya. In fact, in the 1970s it started the first big wave of emigration towards Libya (De Haas 2007b).
7. Through that expression I mean the quarrels among states over the competences and duties in rescuing people, so that each state tends to 'dump' its tasks on other authorities.
8. Harm Greidanus, *Regional Maritime Awareness: Western Indian Ocean, Gulf of Guinea,* www.globalsciencecollaboration.org/public/site/PDFS/Maritime/Greidanus%20H.%20Regional%20Maritime%20Awareness.pdf
9. European Parliamentary Assembly, *Lives Lost in the Mediterranean Sea: Who Is Responsible?* http://assembly.coe.int/ASP/XRef/X2H-DW-XSL.asp?fileid=18095&lang=EN
10. European Parliamentary Assembly, *The Interception and Rescue at Sea of Asylum Seekers, Refugees and Irregular Migrants,* http://assembly.coe.int/ASP/XRef/X2H-DW-XSL.asp?fileid=18006&lang=EN
11. www.frontexit.org
12. 'The ways in which NGOs and humanitarians engage in the governance of migrants and refugees today have changed quite significantly from the kinds of networks of care, self-examination and salvation which Foucault identified with pastoralism . . . the pastoral care of migrants, whether in situations of sanctuary or detention, is not organized as a life-encompassing, permanent activity. Instead, it is a temporary and ad hoc intervention' (Walters, 2011a 158–59].
13. Italian Ministry of Defence, 18 October 2013, http://www.difesa.it/Primo_Piano/Pagine/Via_Mare_Nostrum.aspx
14. 'Passaggio alla gestione ordinaria dei flussi non programmati'. http://www.statoregioni.it/Documenti/DOC_040811_67%20CU%20(P.%2011%20ODG).pdf (accessed March 31).
15. Interview with the Italian seaman, Lampedusa, 30 January 2014.

16. The costs of Mare Nostrum operation are estimated around 10 million euros per month.

17. Interview conducted in Milan on 24 October 2013 with one of the survivors to the shipwreck of 11 October.

18. The surviving Syrians were at first forced to give fingerprints and then were left to escape. In fact, Italy ultimately let the Syrians create their own informal channels for trying to cross the border with Switzerland to go to Germany, since Italy had no interest at all to keep them in the territory; but at the same time it did that on the sly, in order not to be rebuked by the European Union.

19. Encounter with one of the migrants taken from Augusta to northern Europe, Pisa, 18 March 2014.

20. Interview with a migrant coming from Mali and rescued by the Italian Navy. After the arrival at the harbour of Augusta, in Sicily with eighty other people, he was moved to Bologna into a hosting centre (11 March 2014).

21. See the note above.

22. See the note above.

SIX

Unspeakable Maps

Towards a (Non-Cartographic) Counter-Mapping Gaze

November 2011: 'There was a kind of big wave which started to organize the way of leaving, it's [known as] the *harga* [to burn]. This wave comes from Zarzis: the *Quatre Chemins Métro* station in Paris has been well known for twenty years for being the meeting point of people coming from Zarzis. . . . At the beginning it was like wandering, they didn't know anything, they came across many difficulties but they have a savoir-faire and by that they started to occupy the zone of La Villette".[1] Through the voice of M., a Tunisian migrant in Paris, the odd geographies imagined and acted by Tunisian migrants bring out a complex intertwining between invented new cartographies and historical colonial legacies, tracing out unspeakable maps that at least in part unsettle the mapping narrative of migration governance. Through their uneven geographies, Tunisian migrants demonstrated that there could be no real democracy without at the same time the freedom of movement. Once in Europe they moved back and forth across the European space; regional and Eurostar trains became mobile and dispersed borders—some migrants were blocked several times on the trains—and at the same time these were spaces for sneaking off and places for staying. 'This is Europe, this is my Europe': this is the comment made by a Tunisian *harraga*[2] who arrived in Italy in 2011 and who narrated their fragmentary backwards and forwards journeys across Europe by rail. This is not to romanticize practices of migration, but instead to stress migrants' capacity to move across Europe enacting a different pace of mobility than that established by migration policies. Meanwhile, they envisaged their own European geography, corresponding to the experiences they had at the borders: 'Lampedusa is not Europe, and neither is Sicily. Here we are treated as animals. Europe starts northward'.[3] Nevertheless, stressing the spatial redefinition enacted by migrants' geographies does not mean to state the primacy of their European border displacements over the transformations produced by the politics of externalization and the management of migration routes. Rather, in order to come to grips with the reality of a European diffracted space, we need to investigate processes of *border scattering* and border displacements crisscrossing these two mutually responsive spatial practices—migration turmoil and the politics of regional re-bordering.

Taking into account mechanisms of migration governmentality—such as deportations, border controls and temporal politics of migration—this chapter puts into practice a counter-mapping approach, looking at the effects, impacts and resistances engendered both in spaces and on migrants' lives, with a specific focus on the southern shore of the Mediterranean and on the Mediterranean Sea. In the first section there is an overview of the different practices of *dissident cartographies* concerning migration maps, teasing out the main theoretical and political stakes of counter-mapping. After addressing the limits of counter-cartography in the domain of migration, I turn to a *non-cartographic practice of counter-mapping*, taking on counter-mapping as an analytical posture through which to gaze upon and engage with migration and border issues. The non-cartographic practice is staged here through a counter-mapping analysis of the mechanisms of migration management. The goal is not to trace a counter-map of the migration routes across the Mediterranean, but to undo the regime of the visible at play in migration governmentality (Mirzoeff 2011)—and to problematize the possibility of representing migration turbulence through different languages—narratives, cartographies, images. Resisting and opposing the goal of making visible *another map* of the Mediterranean as a space of movements I would insist on the constitutive opacity and elusiveness which characterizes practices of (undocumented) migration. Indeed, this counter-mapping perspective tries to account for knowledges and movements that can neither be integrated and accommodated into a complete story nor, as Chakrabarty suggests, "rendered fully transparent to the gaze of any one particular political philosophy" (Chakrabarty 2000b, 275). Rather, by recalling Deleuze and Guattari's idea that a regime of power is defined by what escapes it, unspeakable migrants' geographies unfold what (in part) escapes the monolingual codification and the mapping gesture itself. By stepping aside from both the register of rights claims and that mapping order, Tunisian migrants interrupted for a period of time the political representation–visual representation that sustains Western modern political spatiality (Deleuze and Guattari 1975).

These *unspeakable maps* are like prismatic devices reflecting and amplifying the spatial-political outcomes of migrants' practices in the Mediterranean, maps that seek to bring to the fore, follow up and foster the spatial persistence and the spatial upheavals enacted by migrants. They are *unspeakable maps* because they don't aim to unfold migrants' strategies at length, producing an overall grid of intelligibility; on the contrary, the counter-mapping approach that I mobilize here brings to the fore the impossibility of making the 'turbulence of migration' (Papastiergiadis 2000) fully readable and visible, both to the governmental mechanisms of capture and to the ordinary codes of political perceptibility (Papadopoulos, Stephenson and Tsianos 2008). There is a margin of 'not-representable' that a counter-mapping perspective should respect. Hence, migration counter-maps actualize themselves in quite paradoxical forms of mapping: in some way, the choice of starting from impossible and fuzzy maps—namely, migration counter-maps—means to gesture towards the Foucaultian approach to power from the perspective of its limits (Foucault 1980a, 1982). Such counter-mapping—as movements acted out of the spotlight of the authorized channels of mobility—resists translation into ordinary codes of visibility. *Unspeakable* is the term I use here for naming such a remnant—that is, all that remains outside the map and outside the existing codes of the political language. Unspeakable migration counter-maps emerge in between the folds of what cannot be said or seen. At the same time, a counter-narrative of migration is also possible only in retaining the *intractability* of some migrants' practices towards being translated into the usual regime of visibility. More than showing the "blind spots" of maps, or their "blank spaces", as Conrad puts it (Conrad 2007), this counter-mapping gaze makes a claim for the constitutive incompleteness and partiality of migration counter-maps. The analytical gesture of counter-mapping strives to foster and push forward the unpredictability and instability of migrants' spatial turmoil, ultimately playing the part of an amplifier of struggles; a kind of magnifying glass with which to shed light on localized and specific movements which dislodge the supposed stability of political concepts and spaces.

UNSPEAKABLE MAPS DO NOT FOLLOW THE FOOTSTEPS OF THE 'ORDER OF BORDERS'

Migrants' geographies undermine the supposed binary relationship between map and territory—or image and reality—highlighting that maps do not (merely) represent but produce spaces and territories and trace or challenge boundaries; they are the temporary result of strugglefields of power relations (Harley 2001; Jacobs 2006). And the reference to the spatial upheavals triggered by Tunisian migrants underlines precisely that, in this case, migrants did not simply move along pre-established borders and paces of mobility, reproducing the existing cartography of the Mediterranean; on the contrary, they performed other geographies of that space. They did not produce other maps, since they did not map space, but they enacted space differently from the cartography of movements traced out by migration policies. Even if they were vanishing or hardly readable, migrants' unspeakable maps have carved out and left long-lasting effects or traces; indeed, enacting the Mediterranean space in a different way from that established by migration policies, and upsetting that cartography, they highlighted the way that practices of migration do not only challenge existing configurations of borders, but also definitively displace that cartography, forcing power to reinvent strategies of bordering. To sum up, migrants' unspeakable maps do not simply act as a counterpoint to the cartography of power, and their disturbing effects cannot be fully recuperated by migration governmentality.

Working within this frame what kind of gaze should we exercise? Where should we turn our attention to? In this chapter I engage in a twofold move, looking simultaneously at the spatial and political upheavals produced by migrants' practices and at the effects that migration governmentality engenders on migrants in their countries of origin and in the Mediterranean space as a space of highly-monitored mobility. The spatial object of this analysis is not restricted to a nation-state, nor does it correspond to areas of controls and selected mobility charted by bilateral agreements; rather it is a space that corresponds not to the Mediterranean area as designed by European policies but to the space coming out from the spatial upheavals generated by the Arab uprisings and practices of migration. Counter-mapping the mechanisms of migration governmentality means at first to shift the gaze from the northern shore of the Mediterranean to the southern one, displacing the European vantage point by drawing the attention to the ways in which migration policies impact upon and are seen from the southern shore. But it is not only a question of decolonizing the gaze and the migration narrative (Le Cour Grandmaison 2005; Mignolo 2009). Rather, the issue should be to unpack certain mechanisms of governmentality, highlighting and challenging the orientation that is implicated in them. For instance, deportations and departures tend to be accounted for and analysed from the European standpoint—no attention is paid to the impact of the deportation regime in the country of origin, while migrants' departures are always considered by deduction, that is to say by focusing on the arrivals on the European coasts; and the impact of border controls cannot really be assessed if we remain on the northern shore, since for instance, as shown in chapter 5, disappearances at sea only become fully perceivable in the countries where lost people come from.

THE CROSS-CUTTING ROUTES OF MIGRATION GOVERNMENT: MOBILIZING (COUNTER) MAPPING TOOLS

Before coming to grips with the topic of migration and border controls I dwell upon the forms of visualization and the cartographic representations of migrations made by international agencies, arguing that they perform and reveal important transformations in the rationale of migration governmentality. In fact, while migration maps have been largely criticized for staging the invasion of the European space by migration flows, representing arrows directed from Africa and Asia towards Europe and reproducing a state-based gaze on migrations (Walters 2009), nowadays the most advanced maps are founded on a quite different blueprint and rationale. In this regard, I take into account the i-Map,[4] the interactive map created by the International Centre for Migration Policy Development (ICMPD) in cooperation with European countries, some third-national countries and agencies like Frontex, Europol, UNHCR and IOM. If, on the one hand, in this interactive map Europe is still envisioned as the space towards which most migrants' routes converge, on the other hand, it is quite noticeable that at the very core of the map there are migrants' routes and not geopolitical national borders. This cartographic shift is a marker of a quite significant shift in the logic of migration management: indeed, the *map-text* is productive and at the same time reveals a subtext formed by mechanisms of migration governmentality, knowledges which have gained the status of sciences and the struggle-field between migrants' practices and techniques of bordering. More than a text, the cartographic tool resembles an inter-text, in which different layers of meaning and intelligibility overlap, and different epistemic orders of truth combine (Deleuze and Guattari 2007).

Moreover, the map as an inter-text refers also to its constitutive heteronomy: far from being a self-standing language, maps intersect and are the result of multiple texts and languages—juridical texts, governmental narratives, geopolitical imaginations and so on. Therefore, in order to be fully readable, maps have to be unpacked through a cross- and in-depth gaze, attentively scrutinizing the composition of different orders of truth and languages. This is, after all, the gesture and the task that Foucault suggests we pursue: 'the aim is not to unfold what is hidden, but rather to make visible precisely what is visible . . . what is so much related to ourselves that, because of that, we cannot perceive' (Foucault 2001f, 540–41). Finally, the inter-textual dimension of maps gains an additional meaning if referred to recent migration maps, where the mechanisms of knowledge production deeply influence the very nature of maps: migration maps reflect the knowledge-based governance upon which migration controls are predicated. That is to say, these maps show in between the folds that migration governmentality is basically grounded on a socialized knowledge—it necessarily needs the combination of a variety of actors, approaches and specific knowledges. This aspect gets a visible application in the crowd maps[5] that are cartographic representations collectively produced, as well as in the i-Map itself, which is likewise a collective cartographic device, constantly updated by different expertises. But most importantly, as explained in the second chapter, the peculiarity of the knowledge-based character of migration governance emerges in the hijacked nature of governmental migration maps. In this way, migration maps follow in the footsteps of migrants' practices to establish a field of visibility and then to realize a cartography which tracks migration for the purpose of outguessing and hampering migrants' future moves, tracing present and expected future routes in order to manage them. Therefore, the spatial snapshots of migrants' movements that the knowledge of migrations captures in order to produce its own cartography is then translated into a temporal map, a map statistically forecasting evolutions and directions of migration routes with the twofold purpose of anticipating/managing movements and providing a specific "economy of gazes" (Azoulay 2008).

In fact, the "cartographic reason" establishes the thresholds of visibility/invisibility of phenomena and subjects (Olsson 2007; Pickles 2004; Turnbull 2000). Maps carve out and inscribe borders, thus telling us as much about spaces as about the tracing of boundaries. In the domain of migration this is translated in the tracing of spaces of governability—that is, migration maps posit how and to what extent people's movements can be managed. And in the entanglement of different regimes of truth and languages and the overlapping of many layers of knowledge, unlike other kinds of maps, migration maps not only stage the narrative through which power situates subjects in space (Crampton and Krigyer 2006), they also complicate such a narrative, intersecting and including different epistemic codes and political sources. Obviously, we should not disregard the 'power of maps' in shaping the real, as well as their double function of legitimation and enforcement of power's strategies (Black 2002; Gregory 2007; Wood 1992). But what it is important to stress at this stage of the analysis is the difference between maps and other texts with respect to the narrative made by governmental actors: the gesture of teasing out the text-map allows us to see conflicting or overlapping knowledges at play in producing the migration regime, unpacking the apparently compact and solid notion of governmentality. To put it differently, a critical reading of the text-map is not equivalent to reading and sustaining the narrative on migrations produced by governmental agencies: could the map be a useful text—not as a neutral tool but as a struggleﬁeld of visibility—to understand the effective functioning of power, both in its actual mechanisms and in its imagined developments? This question would require further investigation that goes beyond the scope of this work; however, what can be suggested here is that the map as an inter-text is situated at the juncture between the effective functioning of power and the way in which the governmental imagination envisages working in the future. In fact, the political dimension of maps does not concern only their direct application into political strategies; it depends also on the regime of visibility that it structures, supporting pre-existing geographies of power and framing specific relationships between subjects and space. Maps, as opposed to other governmental texts on migrations, are operational tools of power which do not only narrate a "governmental phantasy" (Feldman 2011) but also refract the transformations at stake in political technology.

THE GOVERNMENT IS NOT AT/OF THE BORDER: THE I-MAP AND THE REAL TIME UPDATING OF MIGRANTS' ROUTES

Starting from this theoretical background, it becomes easier to assess what transformations in migration governmentality are marked and fostered in the i-Map, as well as in some other recent cartographic productions on migration flows. As mentioned, the first aspect to underline is that what matters in the i-Map is not borders as such so much as migration routes and their orientations: 'migration routes management is reoriented from a focus on a moving front-line to a series of points along an itinerary, redefining "a new architecture of migration management"' (Casas-Cortes, Cobarrubias, and Pickles 2013, 44). Conversely, the traditional mind-set of border thinking vacillates: borders come to coincide with migration routes, on the one hand, and with the spatialities of economic and political intervention, on the other.

What seems to be of relevance in this map is less the directionality of flows than the patterns themselves—that is, the (transnational) spaces that migrants cross, irrespective of the orientation of the arrows. To put it differently, the governmental gaze is turned to migration spaces on the move, instead of focusing on the states 'affected' by migrants' flows; moreover, it is the regionalization of the area to govern which prevails on the state spatiality or on the continental dimension—West African routes, East Mediterranean routes, Central Mediterranean routes and so on. At the same time, the distinction between countries of emigration and countries of immigration is definitively blurred, since what becomes paramount are the spaces of transit and movement and the space of 'spatial insistence' (Sossi 2012) of migrants. Indeed, the towns marked on the maps correspond to the places of transit, departure and arrival, so that the small island of Lampedusa and the city of Oujda are visibly signalled.

Let's try now to take on a map that does not exist yet: a map charting the impacts of developmental politics on non-European countries, and mainly on the neighbourhood countries, which target would-be migrants, stopping them migrating abroad. In a sense, it is a map of would-be migrations that includes the spaces addressed by techniques of bordering. Following this undrawn map, and combining it with the uneven geography of the European space enacted by migrants, the result is that, finally, no map of the European space can be properly traced from the standpoint of the migration issue. On the one hand, migrants' movements (sometimes) succeed in eluding and disrupting the capture of borders and in acting irrespective of the pace of mobility established by migration policies. On the other hand, neighbourhood policies show in a quite glaring way that Europe starts where migrants' journeys are hampered by bordering technologies, breaking up the possibility of maintaining Europe as a spatially well-bounded referent.

However, although the border cannot be the exclusive site of analysis for grasping migration issues in their complexity, it nonetheless remains that the (fragmented) border regime is one of the images through which a coherent European space of free circulation is effectively shaped. These include instruments such as EU cooperation on border patrolling and the standardization of many visa norms and customs controls (Cobarrubias 2009, 70). The spreading of borders is the most direct outcome of migration controls, blocking some routes diverts migrants elsewhere, instantiating another border to cross, and displacing other frontiers: 'Stronger controls at the EU external borders far from solving the problem they seek to solve, actually result in a movement of the border instead' (Rodier 2006, 4). But what is important to remember is that the multiplication of borders generates more than borders, it generates a multiplication (see differentiation of spaces) and essentially it produces distances (De Genova 2013b). Indeed, the tracing of regions and zones where certain politics are implemented—see, for instance, the Economic Trade Areas—and the creation of Regional Protection Programs for refugees (in non-European countries) overlaps with national sovereignties. And in the case of migration governmental strategies, both cartographic representations and discourses envisage a migratory regime grounded on migration patterns and, jointly, on regional/zonal spaces of interventions and monitoring—sub-Saharan migrations, West Mediterranean routes, East African countries.

MAPS OF MIGRATION (IN) CRISIS

Another map acting as a prism of migration governance transformations is the map of the 'Migration Crisis from Libya'.[6] This map is of particular relevance in the frame of this work if we consider that it was launched just at the time of the Libyan conflict and the Arab uprisings. The map provides updated snapshots of the migration crisis triggered by the Libyan war, showing a multilayered surface of visibility on migration factors: border movements, repatriations, humanitarian assistance and people locations are the entries forming the reality of migration crisis as a composite phenomenon to govern. If the i-Map illustrates the government of migration centring on the management of migrants' routes, the 'Migration Crisis' map works through situational risk analysis, creating what I would call migratory compounds that become objects of government. By migratory compounds I mean the outlining of a migration crisis situation whose critical aspect relies on the difficulty of governing mixed migration flows, partitioning them into different mobility profiles. The multiple crises, following the text-map and articulating it with IOM's texts, require as much complexity in the governmental approach to the migrations situation, including all dimensions of migratory crisis—border management, humanitarian interventions and identity checks. Therefore, instead of lighting up and following migrants' movements, the 'Migration Crisis' map spotlights critical border-zones: it locates and marks complex migration phenomena, and migrants' movements come to be included in a much broader object of government, namely, the border-zone of a migration crisis as a complex phenomenon. The production of critical border-zones and the tracking of migration routes are two coexisting mapping rationales, revealing two governmental gazes and operational measures in which the border as a pre-established (geopolitical) line loses its eminence in framing the cartography of migration governability.

In the face of maps like the i-Map and the Migration Crisis map, counter-maps of migrations need to take stock and closely scrutinize the cartographic rationality that they try to counter and challenge. At this stage it is worth briefly taking a step back and casting a glance on the first migration counter-maps produced in the early years of the last decade. These maps are traced against the backdrop of the Fortress Europe imaginary that in the late 1990s and early 2000s was dominant both in critical academic analysis and in the activist debate. Starting from the assumption that mainstream migration maps silence the dramatic consequences of border controls and the regime of detention of migrants, the first migration counter-maps made visible the 'dark side' of migration governmentality, showing a Europe of camps or marking migrants' deaths at the borders.[7] At the same time, other migration counter-maps started to bring attention to the multiplicity of actors and layers of government involved in managing migration and in the chaotic migration regime they generated. In this case, the focus is on the functioning of power itself more than in its violent effects.[8] A third group of counter-cartographies of migrations is formed by those maps that shed light on migrants' practices and on border struggles—namely, the ordinary battle at the borders between migrants and (b)ordering techniques—showing the complex of practices, conflicts and technical tools through which border-zones—for instance, the Straits of Gibraltar—are produced.[9] Then, a huge variety of more artistic counter-mapping practices try to make migrants trace out their own counter-map, hinging on their journey and singular experience of migration.[10] Leaving aside this last group of counter-maps, which does not respond to a traditional cartographic rationale, the common mark of all the other maps that I briefly mention here is their inside position in relation to the cartographic epistemology itself, gesturing towards a strategic counter-use of that epistemic regime. The question to raise thus concerns the effective leeway for engaging in such a counter-use of the cartographic tools. In fact, if we take into account migration maps, two different critical issues overlap. The first one, that critical geographers and NoBorders activists are concerned with, relates to the regime of (in)visibility that official migration maps impose—obscuring the impacts of border controls and portraying migrations as flows threatening the European territory—as well as to the political effects that those maps generate. Thus, taking on this criticism, dissident cartographic practices (Cortes et al. 2008) can effectively use the same language and technical tools for showing *another map*. But as I suggested earlier in this chapter, there is another substantial aspect of maps to sift: the problem of mapping migrations does not concern only the battle *on* and *at* the borders, but the cartographic rationality itself, which fixes and freezes a space of visibility. Indeed, migration implicates more pitfalls and political stakes for the cartographic approach than other topics do. Secondly, the knowledge produced by maps is inevitably coded through the language of representation and signification (King 1996; Papadopoulous, Stephenson and Tsianos 2008), resembling actually what Deleuze and Guattari define as an act of tracing more than mapping: 'the tracing has organized, stabilized, neutralized the multiplicities according to the axes of significance and subjectification belonging to it' (Deleuze and Guattari 2007, 15). For these reasons, the visibility paradigm of cartography, on the one side, and migrants strategically playing with (in)visibility, on the other, clash with each other, with the former encroaching upon the "silent" mode of the latter. Thus, do we need to envisage a counter-narrative of migrations or should we let this map be produced by migrants' spatial practices? What is the usefulness of realizing *another map* and to what extent could it be detrimental for migrants themselves, by revealing their strategies of movement? And, finally, to what degree is a strategic use of mapping tools possible?

'SPACES IN MIGRATION' MAP: TRAVELING WITH MIGRANTS' UNEVEN GEOGRAPHIES

Taking all these critical questions together, the map 'Spazi in migrazione'[11] undertakes a triple displacement of the mapping gaze, troubling the order of the cartographic (in)visibility. First of all, the map tries to account for the spatial upheavals generated in the Mediterranean space by Tunisian migrants who left Tunisia in 2011, and the responses of governmental actors for re-stabilizing migration governability. In this way, the focus shifts from the functioning of power in tracing borders to the spatial effects and the spatial upheavals engendered by migrants, thus subverting the order of space production and suggesting a different gaze on the migration regime: spatial upheavals are triggered by migrants and then national and international actors are forced to respond and reassess their strategy. Secondly, as I sketched, migrants' routes are not made visible but rather are fragmented and translated into a patchy map of the spatial impacts and effects of migrants' practices: the southern European space is troubled and recomposed through the discontinuous presence of migrants and their intermittent (in)visibility and, consequently, uneven mappability. Thus, far from providing a full spectrum of visibility or accounting for all the subjects at stake in that space, the 'Spazi in migrazione' map attentively selects what to highlight and what strategically to leave as unmappable. The third displacement concerns the movement and the location of the mapping gaze: the map is not static but on the move, since it follows the movements of Tunisian migrants, not charting their routes but travelling through the same spatial circuits as migrants to grasp the upheavals they produce. In this way, borders are no longer the landmark through which movements are gauged. By making the gaze travel with migrants' movements, borders are rather what is shaken and transformed by practices of migration. Tunisian migrants, the map tells us, have unified the space between Tunisia and Europe and, through their practices of movement, have wiped out national boundaries. This counter-mapping gesture is notably strengthened by the location of the gaze: turning traditional maps 45 degrees left, the reader is forced to follow the same direction of migrants' movements through a south-north orientation—from Tunisia and Libya towards Europe.

FLIPPING OVER MIGRATION CATEGORIES FROM THE SOUTHERN SHORE

In the second part of this chapter I push forward the non-cartographic counter-mapping approach that I explained and mobilized in chapter 3. However, while there I developed it as an attempt to come to grips with the limit of representation in mapping, here I try to unpack and crack self-standing categories of migration governmentality, engaging in a triple displacement of the gaze. First, turning to the impacts and the effects on spaces and subjects of border controls and migration management; second, highlighting how governmental mechanisms cannot be fully grasped if we remain located on the northern shore of the Mediterranean; and third, peering into the ways in which migrants' turmoil forces national and transnational actors to reinvent a map of migration governability and, more broadly, how migrants' practices make us see the friability of the migration regime.

DEPARTURES

The term 'departure' does not actually pertain to the discursive regime of migration governance, since migrants are counted through the arrivals of people in the hosting countries. Also, when official European statements mention how many people left Tunisia or Morocco, they index departures through the arrivals, without considering those migrants who never arrived. If departures can be deduced only from the arrivals, this is not only because of the clandestine condition of those migrants' departures but also to the irrelevance of that data both for migration policies and for the narrative which replicates the argument of 'migrants' invasion'. The vagueness about departures seems to exclude the very possibility of reframing that notion 'from below' by reversing the gaze and placing it on the southern shore. And this observation leads to a more general argument: the subversion of governmental categories for reading a phenomenon encounters unpassable limits that correspond to the very exclusionary criteria of visibility upon which they are predicated. For instance, asking Tunisian institutions about the number of Tunisian migrants who left in 2011 by boat towards Italy, it emerges immediately that the number could never be reconstructed or tracked down due to the hidden nature of the departures. By definition, no archive of 'illegal' emigration can exist. In order to find the numbers we necessarily need to move to the northern shore, where the Italian government counted 27,000 Tunisian migrants as having arrived on the Italian coasts after the Tunisian revolution. This entails that migrants become visible subjects as and to the extent that they are identifiable. However, this consideration obviously doesn't mean to denounce the failures of the Tunisian government in accounting for migrants who left their country. Instead, the crucial point to bring out is that visa requirements mean that people are de facto forced to devise strategies of invisibility and consequently to disappear—possibly because of shipwrecks but also the sense that they can 'reappear' as subjects only to the extent that they are identified and captured by biometric techniques and identity checks. Nevertheless, they 'reappear' once in the European space as governable and countable migrants in the form of electronic data which tracks their presence on the soil. Thus, the impossibility of reversing the gaze on migrants' departures finally comes to reframe the terms of the problem, highlighting that a counter-mapping approach should insist on emigration as a practice that, in the case of Tunisian migrants, was not claimed as a right but forcibly enacted by them through the revolutionary uprising. Thus, Tunisian migration appears as an extraordinary practice of movement strongly linked to the political uprising in Tunisia. But these practices of migration were at the same time also non-extraordinary if we consider the attitude of Tunisian migrants in leaving the country and the collective dimension of those practices: 'I left just because many friends of mine took that decision, all together; indeed, after the fall of Ben Ali it had become so obvious to take one's own chance and make the harraga—the act of burning frontiers—Europe is so close and I had always desired to go and wander there, also only for a while'.[12] To focus on departures not in terms of numbers but as practices that for many migrants is a flight (Mezzadra 2006) or an act of refusal while for others is related to the desire to change or improve one's own social condition enables us to recall Fanon's observation that 'the first thing the colonial learns is to remain in his place and not overstep its limits' (Fanon 2007, 15). It is precisely when people refuse the injunction to respect the assigned geographical, economic or social place that colonial governmentality can be effectively disrupted. And, in fact, it could be explored how the refusal to comply with the criteria of selected mobility is expressed through migration.

BORDER CROSSING

Both in critical migration studies and in the field of governmental politics, border crossing has a paramount relevance to the extent that all issues concerning people's movements finally coalesce around the nexus of migration-borders. From a genealogical point of view, such a nexus is to be questioned, dismantling the idea of its trans-historical and evident nature (McKeown 2008; Sassen 2006), since it has become the tenet of mobility politics only in the last three decades. But most of all, borders work as vectors of meaning and re-signification for migrations: indeed, practices of migration are framed and approached through the grid of borders, with the latter conceived as sites of control, conflict or negotiation. In this way, border crossing functions as a sort of redux factor packing and narrowing migrations into the very moment and act of crossing frontiers. In this regard, Tunisian migrations displaced the emphasis on borders by putting in place what I call a 'politics of presence' through their unexpected persistence in public spaces and their peripatetic moving (using transport, devising strategies of survival and leaving material traces of their presence but no trace to be counted in statistics). In other words, it refers at once to the concreteness of their *staying there* and to the ephemeral and opaque character of their presence, since their constant need to move away, to change place or to live in public spaces in a concealed way causes us to think about migrations far beyond the act of border crossing. Paris, Rome, Milan, Padova, Bruxelles, Marsiglia and many other European towns are the places where Tunisian migrants tramped and stayed, sometimes moving from one place to the other, or moving by train; however, what is focused on here is not so much the places where Tunisian migrants have lived for months but rather their modalities of staying in those spaces, playing on the edges of invisibility. Thus, mobile sites such as trains have also become places where Tunisian migrants have insisted, as hidden passengers, to travel across Europe over a time span of a few months. Therefore, the very act of *harraga* that Tunisians themselves mention as a brave challenge represents only a delimited moment of migration, since movement itself is acted in multifarious ways, beyond the act of border crossing. For the same reason the borders to be secured are becoming increasingly mobile borders such as trains and ferry-boats.

'WRONG HITS' AND 'FAILING FINGERPRINTS'

The Eurodac database, created in 2003 for fighting against so-called 'asylum shopping'—meaning when a person demands asylum in different European countries[13]—is actually used as a system for also monitoring illegal crossing of European borders: everybody who enters Europe 'illegally' should be fingerprinted and biometric information is sent to the European central database. In this way, all border crossings are in principle detected and stored, and the European perimeter and its borders are thus re-underlined by refraction through the detected presences of the migrants. In other words, digital borders allow the tracing of a map of Europe built on migrants' digital traces. Nevertheless, if we focus on the effective functioning of that data capture and data storage we realize that a highly fragmented map of Europe comes out (Kuster and Tsianos 2013). The data stored in Eurodac does not trace migrants' spatial routes so much as the pace of their movements. Southern European states such as Greece and Italy, the Eurodac annual report states, constantly try to boycott the mechanism by not sending the fingerprint data to the European central unit or sending flawed or fake information.[14] In this way, they boycott the Dublin III logic at its core: in fact, according to the Dublin III regulation, a person who wants to demand asylum needs to do that in the first European country he/she arrives in, and countries that are at the external borders of Europe, like Greece and Italy, are obviously more subject than others to the arrival of third-country nationals. In the case of people stored in Category 2—illegal border crossing—the strategy of southern countries is more ambivalent: by reporting all illegal border crossings they can put pressure on the other member states, claiming the principle of 'burden sharing'; but the number of annual illegal border crossings officially sent by Greece to Eurodac in 2011 corresponds to the average of the weekly illegal crossing at the Greek borders—550 'successful transactions'[15] in Eurodac against 57,000 illegal crossings checked by Frontex at the Greek border. In this way, Greece and Italy can dodge standards and rules about deportations, instead signing bilateral agreements with third countries that include exceptional procedures. In this sense the desirability of Europe and of a European government is constantly challenged by the move of the southern European countries towards East-East or South-South agreements. However, the jamming of Eurodac depends also on the technical failures and incompatibilities between different identification systems that make the "translation" of the national data into the standardized language of Eurodac database an arduous task.

The other logjam that frequently happens in Eurodac consists in migrant's digital traces appearing in a country which does not correspond to the migrant's first point of entry in Europe. In this way the country in charge of processing the demand of asylum is the country that first registered the migrant's border crossing, namely, the entry-point of the migrant in the European space. The mismatches and the jamming of Eurodac result in part from technical failures and incompatibilities between different identification systems that make the "translation" of the national data into the standardized language of Eurodac database an arduous task; and in part they depend on states' refusals to comply with the standards established by the European system of asylum.[16] In this regard, a question intersecting the politics of numbers and the politics of discourses should be raised, since European institutions overtly recognize that logjams very frequently occur (Samers 2004). Thus, how should we conceive a counter-mapping approach to this issue? Indeed, in this case to situate the analysis within the discrepancies between the discursive regime and the effectiveness of the functioning of power is not profitable for outmanoeuvring the mechanisms of migration governmentality. Perhaps the refusal of the European states to send data not only concerning asylum seekers but also illegal border crossings can be understood as a generalized resistance to playing the game of a coordinated government of migration and standardized process of asylum envisaged by the EU.

Which subjects are inside or outside European databases is a question intersecting both the temporality of the identification mechanisms and the individual histories of migrations. In fact, according to the Eurodac regulations, fingerprint data stored from migrants arrested while attempting to cross a European border illegally (category 2) must be deleted after two years, so their presence on the European soil vanishes.[17] At the same time, the data of those who have been returned to their home country are not stored in the system. Thus, deported migrants are not definitively counted and if they come back to Europe demanding asylum (category 1) or arrested during border crossing (category 2), they are categorized as 'new entrances'. More broadly, the mapping mechanism of the European systems of identification results in a scattered map localizing 'illegal' presences at the borders (category 2), irregular migrants living in Europe (category 3) or demands of asylum (category 1). However, a map formed by fixed dots—the punctuated presence of migrants—does not account for the peripatetic patterns of migrants but focuses rather on a moment of the presence of the migrants on European soil and the juridical-political positions assigned to them. Moreover, it is not the actual bodily presence that is of interest to migration governance agencies: rather, what becomes relevant is the attestation of the passage/crossing or presence of that body at some point and in some specific place of the European space (Van der Ploeg 1999).

In December 2011, with the arrival of fifty-three thousand Tunisian and 'Libyan' migrants in Italy, the number of fingerprints stored in Eurodac increased significantly, due to the EU's pressure on Italy. But in this case the politics of numbers went crazy: the discrepancy between the numbers of fingerprints sent to the Eurodac database by Italy and the 'successful transactions' was about twenty-five hundred (53, 000 fingerprints sent against 50,555 correctly processed by the Eurodac unity). Twenty-five hundred people's passage and presence in the European space was lost during the transmission from the national system to the European one. But in addition, the total number of Tunisian and Libyan migrants' fingerprint data taken at the Italian borders appears to be mistaken if we shift the attention to the effective functioning of the *fingerprinting machine*. 'Nobody took my fingerprints in Lampedusa' is the answer that many Tunisians who have now returned to Tunisia gave me, especially those who arrived in Lampedusa in the most crowded months in 2011 when the technical difficulty of identifying people went along with what I previously called a tactic of dis-charge: by letting some of the migrants go undetected on the Italian territory, Italy de facto made it possible to chase them away to France, where indeed the majority were headed.

As Dennis Broeders points out, referring to European identification systems, 'exclusion could take two different contradictory forms: exclusion from registration and documentation and exclusion through documentation and registration' (Broeders 2007, 42). The script of an overall control on people's movements breaks up if we take a step back from the texts and the narrative of power, turning instead to what effectively happened in Lampedusa, since, as Tunisian migrants attested, many of them were not fingerprinted at all. Something escapes and something is allowed to escape. Rephrasing Dennis Broeders' quote on the twofold mechanism of exclusion, it could be argued that both the monitoring systems (radar, satellites) and the techniques of identification work to produce *gaps of visibility* and a *spectrum of non-visible*. And this latter could take two different forms: visibilization through registration and invisibilization because of non-registration or non-monitoring of migrants' passages/presences.

A fundamental counter-mapping gesture consists in challenging the image of Europe as the main destination for migrants from the southern shore of the Mediterranean. Such a gesture mirrors the displacement of frontiers by Tunisian migrants who, as they moved in large numbers towards Libya and Qatar, redrew the map traced by European governments which depicted post-revolution migrations as an exodus towards Europe. Soon after the revolution, due to an awareness of the economic crisis in Europe, Libya became the main economic promise for Tunisia's recovery: according to the Tunisian government's estimations, in 2012, seventy thousand Tunisian workers went back to Libya after the end of the conflict, but since a lot of people did not register at the Tunisian Consulate the actual number is estimated to be around two hundred thousand. However, more recently, with the looming political crisis in Libya and the kidnapping of foreign workers, most have returned to Tunisia and the Tunisian government has stopped any labour placement in Libya. Meanwhile, the high rate of unemployment in Tunisia for skilled people (32 per cent) and the removal, after the fall of Ben Ali, of the restrictions for teachers, engineers and doctors to work abroad, has facilitated an increase in the number of high-skilled Tunisians moving to the Gulf States.[18] Qatar has become a particularly privileged economic partner, signing many economic agreements with Tunisia in different sectors. Looking in the opposite direction—that is, migrants' movements from the South to Tunisia—we should consider the southern Tunisian border: one million people arrived last year fleeing from Libya, and Tunisia decided to leave the border open in order to let people enter.

DEPORTATIONS AND RETURNS

In February 2012, official Italian data recorded thirty-six hundred Tunisian migrants deported in 2011 from Italy to Tunisia, while almost five thousand were deported from France. Seven hundred Tunisians in France came back through the so-called 'voluntary return programs' sponsored by IOM and the *Office Francais de l'Immigration et de l'Integration*, while only around thirty persons have effectively been involved with IOM in Italy. It is harder to find the number of Tunisian migrants who returned by themselves, independent of any governmental program. In principle Tunisian consulates have this data—because most of the Tunisian migrants came without passports and so they need to demand a document to return—but the information is not accessible. Moreover, it is not even completely true that all Tunisians arrived without documents: although this is the case for the majority, some of them brought their passport with them, as Tunisian migrants living in Paris confirmed to me, and consequently their return wouldn't be checked at all. According to Tunisians I met in Tunisia, many migrants who left in 2011 subsequently returned because of the economic crisis in Europe or due to the difficulties they came across in finding a place to live. It can be supposed, listening to some Tunisian voices, that this was a huge number of people, which neither the Tunisian government nor migration agencies 'counted', because they were outside both procedures of deportation and channels of voluntary return. These migrants' informal and ungoverned practices reveal the existence of infra-liminar spaces of migrations that in part dodge and exceed the conditioned mobility regulated by governmental policies. Nevertheless, far from being outside of any constraints, these migrants constantly have to find leeway for moving and living, sometimes enacting discordant practices and strategies of movement which are not immediately graspable by programs like 'migrations for the benefit of all'. For instance, at times they clustered as collectives to better determine strategies of existence and survival, or decide at what point to return home or go elsewhere, trying to escape deportations and moving in a way that does not respond to criteria for 'voluntary return'. Beyond this, we should consider the effects of the deportation regime (De Genova and Peutz 2010) in the countries of origin. What is noticeable is the tendency to study and criticize the politics of deportation mainly in the European or Western space—that is to say, the deportation machine in destination countries, even though there is now a growing literature focusing on deportees (Lecadet 2013; Peutz 2006, 2010; Majidi and Schuster 2013). Beyond the physical, psychological and economic consequences of deportation—families damaged by lack of income and migrants who experience the harsh treatment of police—programs for the assisted return promoted by IOM exclude all migrants who come back as deported, as I explained in chapter 3. In a way, deportation could be seen as a political technology to manage and redistribute undesired moving people (Walters 2002), and starting from that it should be investigated how it functions on the southern shore, that is, in those countries that experience the other side of the same mechanism. For instance, in Tunisia at a governmental level the theme of deportation is somehow eclipsed or neglected: no official discourse has emerged on migrants and their future time in Tunisia in the public debate, even though it is fairly easy to find returned migrants willing to talk about their deportation. Bilateral agreements with Italy and France concerning deportations are negotiated, but then the phenomenon that is presented as something to be governed and fostered is the 'voluntary return', both at a discursive and a practical level, playing on and situating within economic projects for development that reinforce the migration-development nexus.

After migrants are deported to their country of origin, we tend to lose their traces: an analytical scrutiny of governmental mechanisms at a distance necessarily entails a counter-map gesture, which looks at those technologies of government displacing the gaze on the other shore. How are the expulsions of Tunisian citizens seen in the Tunisian political debate? Are they presented as 'clandestine' migrants, caught 'red-handed', or as Tunisian citizens subjected to European migration policies? The first aspect to be noticed is that a debate does not really exist in Tunisia, even after the fall of Ben Ali, concerning the Tunisians who were pushed back from European countries. In other words, despite five thousand Tunisians being expelled from France and almost thirty-four hundred Tunisian citizens from Italy, neither an organized protest nor a political questioning of migration policies has occurred. And the official data of the number of Tunisians deported are taken by the Tunisian government from the European states. The bilateral agreement with Italy signed in April 2011 that stated that all Tunisians who arrived in Italy illegally would be deported directly in a simplified procedure was finally accepted without too many complaints, since the agreement also established the delivering of a temporary permit of six months for all the Tunisians arriving in Italy in 2011, between January 1 and April 5.

But what are the consequences for Tunisian migrants deported to Tunisia? This is a question that can hardly be addressed, since the traces and the traceability of the Tunisian citizens expelled from Europe dissolves quickly, at least since the fall of Ben Ali, because they are no longer being put in jail. However, unlike the sub-Saharan migrants whose journey usually lasts for years before they finally arrive in Europe, in the case of the Tunisians, especially those who left after the revolution, migration requires neither much time nor high economic costs—and some have attempted it many times, or more than once before succeeding. In this sense, expulsion is seen as a kind of false step in the practice of *harraga*. If they are not pushed back immediately after being captured by the European authorities and instead are put into detention centres and then deported, the trace of their presence remains in the European space, both in the Eurodac system—for two years—and in the national databases, for an undetermined time. If we focus on the criticisms of humanitarian agencies and of critical migration scholarship about the politics of deportation, the main target seems to be the noncompliance of North African states with human rights protocols (Ceriani et al. 2009); in this way, the political stakes are narrowed to the question of meeting humanitarian or democratic standards, while the mechanism of deportation in itself is not really contested. The problem is posited in terms of the inhuman treatment that deported migrants risk being subjected to in their country of origin or in countries of transit: North African states are under the demand to keep up with international law on human rights, while European states are blamed for not considering how migrants are treated once they go back in their country of origin. Focusing on sea patrols in the Mediterranean, the eternal debate around the interpretation of Article 33 of the Geneva Convention on the *non-refoulement* (Fischer-Lescano, Lohr and Tohidipur 2009; Lauterpacht and Bethlehem 2003; Liguori 2008; Mitsilegas and Ryan 2010) in which UNHCR obviously plays a pivotal role, finally reinforces the partition between asylum seekers, who have a right to demand protection, and economic migrants. In fact, according to UNHCR, the *refoulement* of migrants at sea is illegal precisely because it hampers migrants' ability to claim asylum and to have access to fair and effective procedures for determining status and protection needs

(UNHCR 2007, 2011). Thus, the floating notion of 'safe country' becomes the tenet through which the politics of deportation is assessed in its legitimacy and fairness, since the deported migrant is considered to be not risking his/her life (Tondini 2010).

In critical migration studies 'deportation' refers mostly to the practices and politics of the expulsion of migrants who have crossed the borders of a nation-state, while the terms 'interception' and 'pushing back' are used for the *refoulements* of migrants at sea. And notably, there are controversies around the term 'deportation', debating whether or not it could be applied to migrants pushed back in high seas, since they have not entered the territory of a nation-state. For this reason, the expression 'technologies of expulsion' enables encompassing an array of ways, conditions and techniques through which migrants are chased away from the territory of a state or from the place where they are (as in the case of interceptions at sea). It is not properly an action of pushing them back, since, for instance, bilateral agreements that Italy signed with some North African countries like Tunisia and Egypt include a clause which requires those states to accept third-country nationals who transited there before arriving in Italy. In this sense, the formula 'readmission agreements' encapsulates this ambiguity, positing expulsions as a return or as a measure adopted by the country of origin or by the country of transit. Despite the fact that the securitization of borders and the management of selected mobility are constantly mobilized in the dialogue between European and African countries, partnerships on mobility are usually enshrined in broader economic agreements on development and free exchange or development and transition to democracy—as in the case of revolutionized Tunisia. After all, the integration of mobility partnerships within the developmental agenda helps in shifting the focus from securitarian concerns to a global approach to human resources—as demanded by Morocco and Tunisia (Rodier 2012).

According to a counter-mapping approach, we have ultimately to take stock of the prominence given to deportations and expulsions as mechanisms to look at for assessing the violence of border controls. In fact, when the number of deportations decreases it does not necessarily correspond to a softer regime of migration governance: for instance, in the two-year period 2006–2008, when Italy decreased considerably the number of deportations to Libya, in the meantime a series of measures were adopted for preventing and blocking migrants' departures from Libya. Nor should we link too quickly deportations and the politics of border spectacle: the first deportations to Libya were enacted on the sly and only a few videos made by activists bear witness to what happened on the island of Lampedusa.[19] Thus, at that time deportations were functional to guarantee that no political or public debate would arise on migration: to keep the island empty instead of producing the spectacle of an invasion.

BORDER CONTROLS

The European agency Frontex is commonly presented, both in critical analyses and in political campaigns, as an exceptional and 'secret' agency playing very often at the limits of the European law itself.[20] And the operative autonomy gained by Frontex in the last five years in relation to the European Union has raised objections from human rights associations and member states, warning the agency of its obligation to respect international and European norms. Nevertheless, it is important to notice that Frontex was created as an operative unit for putting into place—albeit in a political climate of internal quarrels among member states—an integrated border management (IBM). Therefore, the image of Frontex as the official watchdog of the European Union needs to be complicated in order to not reduce Frontex's activity to the securitarian bulwark of Europe. The rubric 'border controls' does not entirely encapsulate the functions and the actions of Frontex, which has rather been working towards and through a pre-emptive logic of risk assessment. Focusing on the Frontex Hermes operation deployed between Tunisia and the island of Lampedusa since 2011, I investigate whether the paradigm of the border spectacle is apt for understanding the rationale and the functioning of Frontex. For this purpose, I shift the attention from an ethnographic analysis of Frontex as an agency with its specific prerogatives to the broader integrated border regime of which it is a component part and that includes not only border patrols and the fight against illegal immigration but also the politics of asylum.

The Hermes joint Frontex operation started on 20 February 2011, as an immediate response to the sudden arrival on the Italian coasts of thousands of Tunisian migrants. The capacity to quickly set up a response was indeed tested by Frontex in 2007, with the start of Rabit, the Rapid Border Intervention Team, deployed in the Aegean Sea to face the Greek 'immigration crisis'. While Hermes is officially supported by many member states, what emerges from the interviews conducted with the Italian military corps in Lampedusa is that apart from the two airplanes provided in rotation by the member countries involved in the operation, all the other means are provided by Italy, which definitively takes over the management of the entire operation. In particular, Guardia di Finanza coordinates the operation, patrolling in the dual role of Frontex operation and Italian military force. The patrolling is conducted not only along the Italian coasts but also on the high sea, reaching the limits of Tunisian territorial waters. In a way, it could be argued that Frontex does not properly exist on the ground, but rather only during operations at sea: what is in place is a set of measures, targets and strategies of action that should be enacted by different national authorities when they patrol as Frontex.

The border regime is characterized by the overlapping of humanitarian borders, security borders and techniques of monitoring and control. In fact, the cooperation agreement signed in September 2012 by Frontex and the European Asylum Support Office can be seen as both a readjustment and a recodification of the border regime, responding to 'complex migration flows coming from North Africa' (IOM 2012b). The cooperation between the two European agencies has produced a substantial blurring of the competences between humanitarian government and border controls, signalling that the former is not the counterpoint or the counterpart of the border regime: rather, the latter continually traces and negotiates its frontiers.

Frontex contributes also to what I call the *stretching of the border*. That expression indicates the displacement of the border from the geopolitical line, multiplying and disseminating it across different sites—before and after that line—and through an array of technologies—monitoring systems, pre-emptive identification mechanisms and biometric techniques. Secondly, along with this spatial disarray it also designates a temporal and conceptual stretching of the border that radically transforms the meaning and the function of the securitarian approach. Frontex is always presented as one of the main actors that fostered the securitization of migrations and asylum (Leonard 2011; Neal 2009) and as the European watchdog, with its team of border guards providing quite an emblematic image of that role. Nevertheless, more attention should be paid to the pre-emptive analysis that Frontex as a research unit produces. The annual risk analysis concerning the 'migration threats' at the external borders of Europe shows that the focus is not only on operations focused at the borders and on a prompt response to migrants' crossing; rather, at the very core there is the production of a real-time picture of what happens at the borders and simultaneously a map that anticipates future migration flows, providing a spectrum of full visibility.[21]

Ultimately, the statistics of illegal border crossing detected by Frontex reveal the peculiar way in which Frontex mixes up border controls and border crossing: indeed, every year the European agency stresses the annual progress in detecting illegal border crossing at the external frontiers of Europe in order to emphasize its increased capacity for capturing the clandestine presences that enter the European space. But what gets lost in this politics of numbers is the not necessarily correspondent relationship between the number of people who crossed the borders and those actually detected by monitoring systems. Moreover, the ability to exercise an efficient monitor-and-capture operation at the borders against irregular migrants is the primary focus, more than succeeding in reducing the total number crossing; indeed, what remains implicit in the analysis is the non-coincidence between detections at the borders and the number crossing. It goes without saying that, in the end, such a discrepancy cannot be measured, since border crossings that are not detected are necessarily also invisible to any count.

ROUTES . . . AND CHANNELS

Migration routes have been notably the main spatial metaphor through which migration management has structured its spatial narrative, at least since the early 1990s. A *struggle over routes* is the image that ultimately emerges from the maps and the analytical reports made by migration agencies like Frontex and ICMPD, on the one hand, showing their frantic chasing after migrations—detecting migrants' routes and apprehending migrants at the borders—and, on the other, the main effect of these actions, which is a substantial reorientation of migrants' movements that tries to find new routes to escape enforced borders. However, despite the primacy of border controls in the logic of migration management—well encapsulated in the disparity in the amount of money that the EU addresses for border controls and in asylum policies[22]—the government of asylum seekers should not be considered a field apart: on the contrary, the interlace between humanitarian and securitarian is precisely the peculiar assemblage and functioning of migration management. Nevertheless, via the vocabulary of the asylum, the spatial metaphor of 'channels' has gained prominence in the migration taxonomy as well as in the governmental rationale concerning migration. In particular, the Syrian conflict has animated a political debate about channels as a mean of spatial relocation. 'Humanitarian channels' is the formula currently employed both by human rights associations and EU politicians and governments for promoting selected facilitated patterns of arrivals in Europe.

THE TIME OF POLITICS

Migration policies frame a complex spatiality of the Mediterranean, creating different channels of access to mobility—visa requirements, free movements, mobility partnership and politics of quota—and consequently producing a forced clandestinity for those who remain out of the selected channels and want to migrate. But the migratory regime also sets the 'temporal pace' of migrations: on the one hand, fixing periods of time in which migrants can be legalized as migrants and then translated into statistics of future expected migrants' flows; and, on the other hand, imposing times of voids and suspension: when undocumented migrants live as invisible presences or when they wait for an indefinite time to get a permit to stay in a certain space. Practices of migration which took place just after the revolutionary uprisings in some way short-circuited the temporal pace of the migration regime, arriving as non-calculated presences and proving to be in a hurry to move on, refusing to be entrapped in Lampedusa. However, this is only one side of the map, since it cannot be overlooked that in April 2011 the Italian government fixed a date limit to give Tunisian migrants a temporary permit, and in May 2011 the European Union proposed a deep revision of the Schengen Treaty concerning the possibility for European countries to reintroduce border controls. That said, it holds true that migrants have succeeded in temporarily disrupting the temporal map configured by governmental politics, while this latter was forced to rearrange itself in the face of the spatial turmoil of migrants' practices. Another way to frame this topic consists in drawing the attention to the temporal narrative that underlies maps and discourses on democracy and migrations in revolutionized spaces. The discourse on liberal conquests and the rule of law in Tunisia is grounded in a temporal logic of stages and on a 'learning democracy' pedagogy that starkly contrasts with the 'disorder' produced by unauthorized movements. The regulated pace of migration politics and the progressive democratic narrative have come along together in the context of the Arab Spring, stressing that the very welcome Spring should be able to translate the messy political unrests into an ordered learning of the rules of democracy to be accomplished step by step. Within such a frame, out-of-place migrations are seen as violating the right pace to get freedom and freedom of movement. Through their practices of movement Tunisian migrants miscalculated for some time both the mechanisms of the selective-and-selected mobility and the logic of a democracy that needed to be progressively learnt: 'everything (and) now' was in the end the goal of their migration translated into words; everything (and) now because democracy is not a set of economic norms, juridical standards and political values to be apprehended but rather, for those migrants, at one with the revolutionary uprisings. In this sense, the shout '*dégage*' against Ben Ali was at the same time a disavowal of the legitimacy of any form of government over lives (Sossi 2012).

The geographies that Tunisian migrants traced out in wandering across the European space definitively undo the idea of a coherent and stable migration regime, since, as these counter-mapping gestures have shown, the existence of a political technology governing migrations is constantly undone by migrants' presence in space. In other words, migrants' practices work as resistances against governmental mechanisms of capture and border controls that try to discipline, filter and bridle practices of movement. In a nutshell, a counter-mapping approach on migration governmentality allows us to take on 'resistances as a chemical catalyser, enabling us to underline power relations' (Foucault 1982). To put it in cartographic terms, instead of taking for granted the codes of political grammar, the counter-mapping gesture consists in showing how migrations (sometimes) actualize discordant practices of freedom in relation to the existing cartographic regimes of borders. In fact, migrants' strategies are within the same space and map that produce and capture them as 'illegal migrants', while at the same time they also shake the legitimacy and the tenability of those cartographies. In other words, they disrupt the existing geopolitical map of Europe and, simultaneously, open new spaces: these are movements that in governmental maps are not silenced or invisible but, more radically, are not expected to be there. In some way, through their uneven geographies, Tunisian migrants broke up the official narrative of the Mediterranean as a space of free circulation—showing the deep asymmetry between the two shores—and enacted an unpredictable freedom, crippling for a period of time the mechanisms of selected mobility.

NOTES

1. 'Vieni qui al Metro Quatre Chemin', Interview conducted by Federica Sossi and Martina Tazzioli with a group of Tunisians living in Paris (November 2011) in Sossi, 2012.
2. 'Harraga' designates both the act of burning the frontiers (moving 'illegally') and the act of burning identity documents. Thus, 'harraga' is also a term which many Tunisian migrants call themselves.
3. These words were frequently repeated by Tunisian migrants blocked on the tiny island of Lampedusa in February 2011.
4. International Centre for Migration Policy Development, http://www.icmpd.org/i-Map.1623.0.html
5. Crowd maps are interactive collective maps built on the model of open platforms, which allow people to constantly update a certain map concerning a specific phenomenon. See for instance Crowdmap: https://crowdmap.com/welcome
6. International Organization for Migration, 'Migration Crisis from Libya', http://www.migration-crisis.com/libya/
7. United against Racism, www.unitedagainstracism.org/pages/map_Fortress Europe_OWNI.htm; Migreurop, www.migreurop.org/IMG/jpg/map_18-1_L_Europe_des_camps_2011_v9_FR.jpg; Le monde diplomatique, http://blog.mondediplo.net/2010-06-01-Les-camps-d-etrangers-symbole-d-une-politique
8. Transit Migration, http://www.transitmigration.org/migmap/
9. Hackitectura, http://mcs.hackitectura.net/tiki-browse_image.php?imageId=580
10. Harraguantanamo, http://www.youtube.com/watch?v=him1yL5YTcw
11. The map is included in Glenda Garelli, Federica Sossi and Martina Tazzioli, eds., *Spaces in Migration: Postcards of a Revolution* (London: Pavement Books, 2013).
12. Interview with a Tunisian migrant (in the Tunisian city of Zarzis, July 2012) who returned to Tunisia.
13. 'A system known as Eurodac is hereby established, the purpose of which shall be to assist in determining which Member State is to be responsible pursuant to the Dublin Convention for examining an application for asylum lodged in a Member State, and otherwise to facilitate the application of the Dublin Convention under the conditions set out in this Regulation' Council Regulation (EC) No 2725/2000 concerning the establishment of 'Eurodac'.
14. Eur-Lex, http://eur-lex.europa.eu/LexUriServ/LexUriServ.do?uri=COM:2012:0533:FIN:EN:PDF
15. The Eurodac regulation uses this expression to indicate 'a transaction which has been correctly processed by the Central Unit, without rejection due to a data validation issue, fingerprint errors or insufficient quality'. Eurodac, Annual Report, COM (2012), 533.

16. A wrong hit occurs when a third-country national lodges an asylum application in a member state, whose authorities take his/her fingerprints. While those fingerprints are still waiting to be transmitted to the Central Unit, the same person could already present him/herself in another member state and ask again for asylum. A missed hit occurs if a third-country national is apprehended in connection with an irregular border and while those fingerprints are still waiting to be transmitted to the Central Unit the same person already presents him/herself in another member state and lodges an asylum application.

17. Regarding 'digital deportability', Kuster and Tsianos convincingly argue that 'it is the result of the permeability of Europe's borders, making deportation at any given moment a constant threat within the slick space of the data flow. It is not the migrants themselves who circulate here, but rather the "embodied identity of migration," as the sum of their data doubles' (Kuster and Tsianos 2013, 1).

18. The official number of Tunisians working in the Gulf States is given by Tunisian job agencies that select high-skilled Tunisians to work in the Gulf States. Between Qatar, Saudi Arabia, Oman, Kuwait and Arab Emirates the number in 2013 was of fifty-eight thousand five hundred Tunisian citizens.

19. The first deportations from Italy to Libya were made in 2004, and in the period 2004–2006 more than thirty-five hundred have been deported. But the first interception at sea took place in May 2009.

20. Frontex, http://www.frontex.europa.eu/assets/Attachments_News/119780.pdf

21. The Frontex Situation Centre has the task of providing a constantly updated picture, as near to real time as possible, of Europe's external borders and migration situation. (http://www.frontex.europa.eu/assets/Publications/General/Frontex_Brochure.pdf).

22. The EU Solidarity and Management of Migration Flows Programme (SOLID) in the period 2007–2013 allocated 4 billion of euros to support member states' activities on asylum, migration return and border controls. The percentage designated for external border funds was the 46 per cent and 16 per cent for return funds, while only 17 per cent was for the refugee fund and 21 per cent for the integration fund. And this discrepancy is even more relevant in the case of EU member states that are subjected to massive migrant arrivals: for instance, Italy received 36 million euros as refugee fund and 250 as border fund.

Conclusion

On 15 July 2014, while this work was being written, a *twofold spatial upheaval*—migrations and the Arab uprisings—was underway in the Mediterranean. The date of these conclusions marks three and a half years since the outbreak of the Tunisian revolution. Writing in the turmoil of these events imposes the limit of not being able to envisage the future developments of events underway, and of not being at sufficient distance to take stock of these spatial reshapings. Nevertheless, this work has tried to come to grips with the contingency of the present context and its changing conditions, resisting the temptation to judge the failures or successes of revolutionary movements. The dates of the snapshots throughout the book correspond to relevant issues/transformations or to moments of rupture that took place during the Tunisian twofold upheaval and also later, as it resounded in Europe across different spaces and times. Moreover, they also indicate the frantic effort to keep up with the hectic reassemblages of migration policies: indeed, also at the time of writing these conclusions new bilateral agreements are being established and new mechanisms of capture are being implemented for responding to the 'unruly mobility'. By dating the conclusion of this work July 2014 I've chosen a moment to situate the gaze on the present, since the conclusion of this work does not correspond to the completion of the turmoil, despite media and political analyses depicting the current context as a post-revolutionary stage. However, the resistance to speak in terms of 'post' and to imposing the temporality of the events on political uprisings should not preclude taking into account the transformed social and political context in revolutionized Tunisia. The ongoing political instability has transformed revolutionary uprisings into more diffused local fights that bring to the fore the deep distrust towards any form of representative politics, as the general strikes and 'exoduses' confirm.[1]

Here I pinpoint some of the most relevant issues that were raised in the previous chapters and that require development or that open the ground for further interrogations. What emerges is not a narrative but a texture, currently in the making, that attempts to retrace and follow up on the main concerns that the spatial upheavals have generated, highlighted or exploded.

POLITICS OF PERCEPTIBILITY

The non-cartographic counter-mapping gaze on migration governmentality has allowed us to see the limits of the politics of representation that underlie the activity of mapping. A salient issue emerging from this counter-mapping approach consists in the difference between a politics of visibility and a politics of perceptibility. Visual approaches hinge on the power of images to produce political spaces and boundaries and to strategically overturn the directions of power over knowledge and image production (Mirzoeff 2011). Many of these visual approaches explicitly draw on Jacques Rancière's thesis on the disruption of the thresholds of visibility in order to create visible phenomena, practices and subjects that remain *out of the spotlight* and *off the map*. In this way, the primacy of the visible and the struggle over (in)visibility are posited as the main axes upon which political subjectivities can be refashioned. Therefore, the politics of perceptibility—as an interrogation disrupting the thresholds of the tolerability of power—is flattened and conflates into a politics of visibility. In other words, the assumption that sustains a visual approach to migration is that the disruption of the codes of visibility would also engender a transformation of the thresholds of perceptibility and activate a transformative politics. If it is indisputable that the growing production of images and critical maps on migrations has brought to the surface new fields of visibility, then the effects of this visibility need to be deeply questioned. Actually, the regime of perceptibility encompasses all the ways through which 'noisy' or silent practices become part of our political horizon and space. More than bringing out into the existing political space these noisy or silent movements, we should pay attention to the way in which they upset such a space. The practice of 'mapping otherwise' is neither a condition nor a guarantee of shaking the thresholds of what is acceptable or not; instead, a politics of perceptibility gestures towards a transformation in the practical and critical attitude in the face of power. In this regard, migrants' spatial upheavals that have spread across the Mediterranean have deeply destabilized the vocabulary of political recognition through which migrants are also made visible on the map. Anesthetization and acceptability: these are two mutually related effects of the proliferation of visual products on migration and the mechanisms of bordering. While acceptability represents what Foucault has defined as the threshold of intolerability of power which makes, at some point, people rise up, anesthetization is the effect against which the work of critique acts. However, when the disruption of the thresholds of acceptability is articulated by the act of 'making visible', as in migration, the risk is paradoxically that the thresholds will be raised, producing a sort of anesthetizing effect that prevents any transformation and active engagement (McLagan and Yates 2012).

MIGRANT GEOGRAPHIES AND VANISHING MAPS

Tunisian migrants have envisaged and practiced an erratic and uneven mobility, displacing the spatial order of the Mediterranean and its implicit directionalities—which see migrants going northward and democracy exported southward. Moreover, through their practices of movement, they have also envisaged the European space as if it were at hand, in contrast with the distance and the asymmetric mobility between the two shores of the Mediterranean that migration policies put into place. Their choosing where to go once they arrived in Europe, refusing to stay in Italy and showing a considerable ability to move around despite their unauthorized condition, is the most tangible mark of how they enacted a different map from the governmental cartographies. The imaginary of Europe that Tunisian migrants traced out is framed by the times of their journeys and by their strategies of migration. Some of their journeys were blocked in rail stations or proceeded by local trains; others involved squatting in buildings until being evicted. Some Tunisians arrived in Belgium or in the Netherlands from Lampedusa and were subsequently pushed back to Italy; sometimes they moved very quickly, while other times they were forced into indefinite stays. For those who had contacts in France, Paris was close, and for others, so too was Sicily, while on the contrary the distance between Rome and Bologna could be significant, given the risk of being arrested on the train by the police. What emerges is a European geography reshaped by the effective ways and possibilities of acting in space: migrants' erratic presence in the European space and the stories of the others who migrated before them combine to shape a fragmented *Europe in migration* whose borders and distances reflect the effective time of migrants' moving and staying. The democratic spatial orientation of the Mediterranean has been definitively shaken by Tunisian migrants who arrived in Europe not with the aim of discovering democracy but as the children of the Tunisian revolution. Through their erratic movements, Tunisian migrants put into place *other maps* of the European space, stretching distances and borders through their temporal pace made of movements and rests. Concurrently, the relevance of cities and places was redefined by their knowledge and imaginary of those spaces, as well as by the possibility to 'make use' of them. European geographic landmarks were positioned by migrants according to the contacts they had or the life projects they were able to realize there: Parc de la Villette in Paris as a place well known to Tunisians for ages; Norway as a country in which to go unnoticed; Switzerland as a state to apply to for asylum; the occupations in Padua, Paris and Marseille as places to find shelter.

However, three and a half years after the *twofold spatial upheaval* began, the vanishing character of migrants' maps cannot pass unnoticed. In some way, the multiple cartographies of Europe performed by migrants resemble the elusive pace of migrants: putting them into resonance with migration governmental maps, they show the hindrances and impacts of borders on migrants' movements. But at the same time, migration's uneven maps emerge from the cunning strategies through which migrants exceed the pace of mobility set by migration policies. Migrants' maps break up the consistency of Europe as a space with one single temporality of movement and one conception of space, as 'an object constituted discursively' (Chakrabarty 2000a) and through cartographic imagination (Sakai 1998). Following this point, it could be argued that migrations bring out the fundamentally fragmented and composite dimension of Europe, in contrast with the supposed bounded and self-outstanding political community of the nation (Anderson 1990)[2]: the enacted uneven geographies of migrants in Europe show that the imagined Europe is a protean and many-sided cartography, formed of heterogeneous—and conflicting—narratives.

Did migrants' geographies put into place discordant practices of freedom that do not reproduce what is already there? This is a crucial question, given that what seems to emerge three years later is precisely the difficulty of finding some traces of the collective struggles migrants organized and of the disorder they generated in the political geometries of the European space of free circulation. Visiting the most conflictive sites of the 'Tunisian revolution in Europe' three years later, I noticed the disappearance of the most important political markers of their presence—such as political collectives or occupations of buildings in European cities. In Paris, Milan, Rome, Brussels and Marseille, places where Tunisian migrants stayed for months, few traces remain of their spatial upheavals. In fact many have gone back to Tunisia, while those who remained on the northern shore are scattered all over the European space. However, the point here is not to romanticize the elusive and vanishing dimension of migrants' movements. Rather, this issue should be posed in terms of the possibility of envisaging other maps and alternative common vocabularies of struggles that are not grounded on nation-based scripts. All these things considered, I suggest shifting attention to the circulation of languages and practices across different spaces, and to what I call the deferred bounce-effects of the struggles. Despite the short duration of many migrant struggles and the uneven strategies of (in)visibility they perform, if we turn to the resonances of those practices as deferred in space and in time, a quite different landscape emerges. This can be formulated in two questions: how did some claims and struggles circulate in multiple sites? As shown in chapter 3, we can shift the focus to migrant struggles taking place at a temporal distance, looking at the reverberations and the recalling of existing or concluded struggles, like in the case of the *Lampedusa in Hamburg* collective that was put into place recalling the *Collectif des Tunisiens de Lampedusa à Paris*.

Moreover, along with this spatial and temporal dislocation of the gaze, I suggest that migrant struggles cannot be narrowed to those in which people are involved only inasmuch as they are labelled 'migrants': indeed, a fundamental gesture consists in looking at migrants involved in struggles not as migrants. This is also related to the refusal of collapsing all conflicting sites and mechanisms of identification and subjectivation into the referent of 'migrant', as if this is the primary and overwhelming field defining its figure of subjectivity. Moreover, as stressed above in the chapters, there are many life and labour conditions, mechanisms of exploitation and power relations in which people are involved, shaped and captured that go beyond the label/status of migrant. For instance, the struggles for housing that are mushrooming all over Europe see migrants and other city residents involved in the same battle, starting from a common condition that cuts across the juridical status.

Indeed, governmentality as an operative methodological tool precisely enables us to see that it is not in terms of binary relations that the strugglefield of powers/resistances can be framed, revealing that the issue is not to endorse one of the two poles of the battle—visibility vs. invisibility, identities established by power vs. autonomous subjectivity. Rather, it is a question of playing within and along the subtle ambivalences of processes of subjectivation through which practices of mobility are translated into migrations. The ambivalent ways in which migrants grapple with mobility profiles and identities show that a critical analysis should retrace the emergence of a certain field of government and illustrate how local tactical elements are situated in broader strategies. A concrete example comes from the multiplication of mobility profiles through which migration governance partitions migrants, hampering in this way any possible unity of struggle. The importance of pluralizing the catchword of migration is not in itself a political defeat to the migration regime; instead, it depends on the very logic upon which migration governmentality is predicated: migrants and asylum seekers who get muddled in the mechanism of partitioning cast light on the inadequacy of migration categories in encompassing all forms of mobility. However, different migrant conditions are not reduced to the juridical status and the mobility profiles through which migrants are fixed to an identity. In fact, juridical categories need to be confronted and articulated with the regime of labour and with racializing mechanisms. What emerges is a complex picture, in which binary distinctions and well-defined subjectivities are superseded by a more blurred cleavage of inclusion through exploitation of (some) migrants in the labour economy. Therefore, an exclusive focus on the epistemic and juridical proliferation of migration profiles does not make clear the functioning of mechanisms of inclusion and exclusion—and most of all, of 'exclusion through inclusion' (De Genova 2013a)—which are at stake beyond the smooth space of subject-positions depicted by governmental agencies and juridical knowledge.

Confronted with that, categories and profiles ultimately explode into an entangled set of effective conditions and different ways of being partitioned, bordered and captured—labour contracts, resident permits, regularization programs and so forth.

The script of the 'interruption' that I mobilized in chapter 1 seems to overcome the supposed antinomy between constituent and destituent processes: indeed, the former emerges in excess of the immediate goals and the explicit claims, and a collective dimension is eventually the outcome of reiterated practices of struggle circulating across space. The interruption of power's mechanisms—that taken per se could result in a particular disruptive moment—by breaking the chain of political signifiers, or disrupting the functioning of bordering techniques, could trigger the spreading of shared practices that open spaces of movements and free stay that are not expected in governmental cartographies. 'Border interruptions' are disruptive moments and practices that shake the readability of maps and that trouble the temporal order of the progressive democratic transition. In particular, border interruptions refer to struggles and practices that take place around border spaces and zones—well beyond the border line—or that contribute to reshape the location and the function of boundaries, challenging and interrupting their existence and their selective mechanisms. Indeed, as illustrated in chapter 6, sometimes migrants' geographies (and not every migration in itself) break up the longstanding association of political representation and cartographic representation.

THE REFUSAL TO STAY IN ONE'S OWN PLACE: MIGRANTS' UPHEAVALS IN POSTCOLONIAL SPACES

Resistances enacted by refugees and their strategic positive stories; Tunisian migrants who refused to play along with the stages of the democratic transition and enacted their freedom of movement, pushing the impact of the Arab uprisings to the northern shore of the Mediterranean; rejected refugees at Choucha camp that undermined the principles of the politics of asylum, demanding protection as having escaped from Libya and from a war started by Western countries. These three episodes confront us with migrants' refusal to stay in their expected place, namely, the place established by the articulation of migration policies and economic factors. However, 'place' does not stand here simply for being fixed in space: rather, it indicates the times and the conditions at which migrants are supposed to move and stay. In particular, in the case of the Tunisian migrants, such a refusal was actualized through their unexpected presence on European soil, troubling the narrative of freedom and democracy conquered through progressive steps. As Fanon remarks, the refusal to stay in one's own place, conceived both in a spatial and in a social sense, is one of the main forms of resistance through which the colonized unsettled the colonial order (Fanon 2007). And this provides us with a lens to grasp the *intractability* of migrant struggles when confronted with the codes of representation and citizenship. Indeed, what characterizes these movements is, I suggest, the way in which they interrupted the pace of the politics of mobility for some time; not so much by making claims but by enacting practices of freedom that exceeded the 'tractability' and governability of their conduct by migration policies.

UNSETTLING THE METHODOLOGICAL EUROPEANISM

The travelled and uneven maps of Europe enacted by Tunisian migrants, and the political uprisings on the southern shore of the Mediterranean, unsettled the *methodological Europeanism* which underpins both migration studies and the analyses of the Arab Spring. In fact, migration studies tend to posit Europe as the master signifier and the spatial referent of migrations, postulating a strong desirability of that space from the migrants' standpoint. Similarly, political analyses on the Arab revolutions read those events in the light of the Western road to democracy, by iterating the script of secularization and democratization in the Arab countries according to the logic of the 'not yet', overlooking the political practices taking place on the southern shore. The growing centrality gained in the political debate by the Euro-Mediterranean signifier—as a space of free exchange and free mobility—contributed to taming and disciplining those upheavals, encoding them into a narrative mastered by the prefix 'Euro': indeed, through the cultural and geographical medium of the Mediterranean, the spatial disarray triggered by migrants' movements and political uprisings were rebalanced into that supposed shared space driven by a European matrix. In this way the silent referent of Europe is surreptitiously reintroduced through the Euro-Mediterranean frame.

In this regard, the Foucaultian perspective of a history of our present comes to be strongly complicated in the face of the *twofold spatial upheaval*. Addressing both migrants' practices and the political turmoil that burst on the southern shore, it becomes necessary to pluralize our present, paying attention to the different ways of practicing the space and to the asymmetries in terms of possibility to move and stay that carve out the space of free circulation. The tracing of other maps of Europe migrants' geographies unsettles the very idea of a unique space in which people move or stay in different ways. Therefore, a history of the present requires at the same time a pluralization of its terms—'histories' and 'present(s)'—and in turn such a gesture fractures the consistency of a given homogeneous space to investigate. The second displacement concerns the spatial coordinates we take for granted: the pronoun 'our' presupposed in many analyses that interrogate the bank-effects of the Arab revolutions (what impact did they produce on 'our space'?) comes to be deeply questioned. Does it refer to a contemporaneity that encompasses many spaces but keeps alive the blueprint of European political modernity? Or rather does it result from the complication of such a spatial/temporal baseline—through the resonance of multiple events? I suggest that it concerns the refusal to immediately translate political practices and claims into the blueprint of political modernity. A gaze on the *Tunisian upheaval in Europe* enables us to envisage the present European space as eminently written and unsettled also from the 'outside' and as a space that is actually formed by different enacted geographies—how the same space is crossed and lived differently.

MIGRATION CRISIS, CRISIS AS BORDERING

The conquest of democracy in the Arab countries is the main motif through which the Arab uprisings were narrated from the northern shore of the Mediterranean. Nevertheless, as this book has shown, the paradigm of the road to democracy has been quickly reversed into the framework of the crisis. In particular, in revolutionized Tunisia migration as a troubling factor was associated with the overwhelming catchword of the 'crisis' that the Libyan conflict generated, spreading far beyond Libyan borders and designating multiple sites of crisis. Thus, 'crisis'—conceived both as a discourse and as set of governmental strategies—works as a re-bordering technology that traces emergency areas, border zones and new spatial economies. On the one hand, the politics of 'migration (in) crisis' mobilized in the aftermath of the Libyan conflict has in part redefined the terms of migration governmentality and of the politics of asylum. On the other hand, it has been characterized by the persistence of the horizon of the crisis: the issue does not seem to be how to step out of the crisis, but rather how to revise the politics of mobility, taking the crisis as the insuperable baseline. It is worth stressing how re-bordering processes triggered by the crisis restarted processes of primitive accumulation: as Sandro Mezzadra persuasively argues, during periods of (economic) crisis 'primitive accumulation is continuously restaged as a problem that interrupts the historical linearity of development, in particular in periods of crisis, when capital deploys—in the face of specific limits—its nature of permanent revolution' (Mezzadra 2014, 25). However, if one stops the analysis here it might appear that a circular dynamic between power-resistances is at play, according to which migration policies reassess their strategies in the light of political transformations and migration turbulence. Instead, that supposed circular mechanism is what must be challenged, highlighting the disconnections and the breaking points that migration produces inside. In this regard, as analysed in chapter 4, practices of migration do not only resist the script of the crisis but also produce it, making governmental partitioning spin freely and forcing power to rearrange its strategy as a strategy of crisis.

REORIENTED MIGRATIONS ACROSS THE TWO SHORES OF THE MEDITERRANEAN

This final area of focus points more towards further developments of this analysis of the *twofold spatial upheaval,* undertaking a counter-mapping move. Indeed, the spatial reshaping engendered by the Arab uprisings in the Mediterranean should be articulated today with the reorientation of mobility caused by the economic crisis in Europe. The geographies of movements between the two shores of the Mediterranean have actually been complicated by the increasing economic backlash. The European space itself emerges as deeply fragmented and characterized by the reshuffling of inequalities among European citizens, giving rise to processes of migrantization of people from southern Europe. In particular, the reorientation of migration across the two shores is a visible marker of the effects of that articulation, and it shows us the increasing provincialization of Europe as a space of immigration and as the 'sovereign subject of all histories' (Chakrabarty 2000a; see also Chen 2010). In this regard, as illustrated in chapter 4, the Tunisian revolutionized space is becoming a *migration cluster*. Far from being exclusively a country of emigration towards Europe, migration patterns that increased after the outbreak of the revolution tell us another story: migrant workers, asylum seekers and transit migration coming from other Maghreb countries, from Libya or from the sub-Sahara region. With the revolution and the growing economic crisis that is affecting Europe, this scenario of multi-directional migration has gained more consistency: the Libyan war caused the arrival of hundreds of thousands of third-country nationals as never before and the temporary return of many Tunisians working in Libya. But the intensification of the rate of unemployment in Tunisia and in Europe has hugely increased the number of Tunisians moving to the Gulf States to work as skilled or unskilled migrants, also as a result of the closer economic and political cooperation by Tunisia with those countries.[3] Nevertheless, the reorientation of migrations does not concern only Tunisians or people from other African countries. In fact, in the last five years the number of European citizens—especially Italians and French—who have migrated to Tunisia to find a job has considerably increased. At the same time, the cheap costs of the labour force is pushing many European industries to externalize their production in Tunisia. The result is a new orientation of migrations that would seem quite unexpected to the eyes of European readers: young unemployed Italians, some of them also with a university degree, have chosen to migrate to Tunis to live and work, profiting from the low cost of living in Tunisia.

The articulation between the economic crisis and the *twofold spatial upheaval* forces us to reiterate the question 'who is a migrant, here and now?' and to critically interrogate how some practices of mobility are seen as practices of freedom and others as migration. The interferences between migrants' upheavals and the Arab uprisings suggests a take on migration as a contested strugglefield and a need to challenge, rework and unravel the boundaries of migration as a self-standing field of research. To shift the gaze to the reorientation of migration across the Mediterranean and, in particular, to European citizens moving to Maghreb to find a job is part of a decolonizing gesture that displaces Europe as the most desirable space for living and as the primary migration destination. Jointly, instead of assuming a spatial reference—Europe, the Mediterranean—as a surface, such a gaze starts from practices of movements to see how existing cartographies are troubled and transformed, and also how new spaces of mobility and stay are created.

MAKE SPACE, NOT BORDERS: CONSTITUENT PROCESSES STARTING FROM THE BORDERS OF EUROPE

If one of the issues of this book consists in the refusal to reproduce the partition between asylum seekers and migrants, and more broadly between 'legitimated' migrant presences—as migrants in need of protection—and economic migrants, this must not prevent us from doing a thorough analysis concerning the different political and subjective conditions of migrants' departures. Indeed, the spatial upheavals produced by Tunisian migrants in Europe in 2011 and in 2012 and the current presence on the territory of people escaping Libya and the Syrian conflict cannot be equalized. The mass exodus of families from Syria, as well as the many war conflicts in Mali, Eritrea, Iraq and Palestine, show us a geopolitical context that is undergoing deep transformations and that cannot be approached only through the lens of migration but requires an articulation of political, economic, religious and geopolitical factors.

* * *

Milan, central train station, July 2014: hundreds of Syrians—women, children and men—arrive every day at the station. They have not been fingerprinted by the Italian police, violating the Dublin III regulation, but have been allowed to move from Sicily, where they arrived through the military navies of Mare Nostrum, and to travel to Milan by train. Then, from Milan, they will try to cross the Swiss or the French border to get to Germany and Sweden. Syrians wait at the train stations to be placed in one of the many temporary hosting centres, staying there only a few days, just the amount of time necessary for organizing their 'illegal' border crossing to Switzerland and France. The 'clandestine disobedience' performed by Italy[4] that does not comply with European rules is not made in the name of people's right to choose the place to stay and to claim asylum, but rather in the hope of chasing them out of Italy. The massive presence of Syrians at the train station in Milan, like in many other Italian towns, cannot pass unnoticed to anyone who lives or goes through those places. However, despite their visible presence, there seem to be invisible or unnoticed channels of people moving northward across Italy, since they do not trouble the ordinary perceptions of citizens in any way: they are visibly there, but the scene has become something too common to notice.

Nevertheless, what stands out both through the Tunisians' upheavals, and through the unnoticed channels of Syrian families, is the political and juridical inadequacy of Europe in keeping up with a practice of hosting that does not consider migrants as 'non-citizens' to tolerate on the territory, but as presences through which the European space itself needs to be reshaped beyond citizenship and foreignness. *Make space, not borders* is the actual ethical and political stance upon which the challenge to the border regime and the existing spaces of governmentality hinges. Also before being afforded the right to stay, it is a question of struggling for a right to have a space and to have chosen it; it means striving for producing common spaces in which no presence and movement is out of place or unauthorized, and refusing to accept that some categories of people need to ask permission for travelling or staying in a space. This is something that pertains neither to the rights violations nor to the incompliancy of states towards international law, since the unequal access to space is precisely what is sanctioned by the states' system and by migration policies. Thus, *make space, not borders* cannot be enacted unless it is through constituent processes that rethink and reshape the European space starting from its borders and from its non-citizens.

NOTES

1. For instance, in 2012 many general strikes took place in Tunisian towns. One of the most important was in the city of Siliana, 20 November 2012, when the residents decided to start a strike to protest against the high percentage of unemployed and the unequal distribution of economic investment between the inner regions of Tunisia and the regions of the coast. The residents also left the city, in a kind of symbolic exodus, in protest against the lack of attention and investment on the part of the government. Another significant protest happened in the village of El-Redeyef, in December 2012: also in that case the residents opted for a symbolic exodus towards Algeria, protesting their exclusion from the governmental programs of employment. A similar movement took place in the city of Ghardimaou in January 2013.

2. On this point, see Partha Chatterjee and his criticism of Benedict Anderson's theory of nationalism as an imagined community: in fact, according to Chatterjee, while, on the one hand, Anderson 'refuses to define a nation by a set of external and abstract criteria', on the other, he conceives of imagined communities in terms of a

'modular character' of culture (Chattejee 1986, 19–22). Also Talal Asad, in *Formations of the Secular*, questions Anderson's idea of nation as an 'imagined community' that is grounded in homogeneous time (Asad 2003, 2).

3. The number of skilled Tunisian citizens who moved to the Gulf States has almost doubled between 2010 and 2013 (1,410 in 2010 and 2,780 in 2013). This is in part an effect of the Tunisian revolution—since after the fall of Ben Ali Tunisia removed the cap on teachers and other skilled professionals who want to emigrate—but it was most of all the outcome of the economic crisis in Europe, which, on the one hand, impacted on the Tunisian economy—especially on skilled jobs—and, on the other, deterred Tunisian people from moving to Europe to find a job.

4. Actually, not all migrants are allowed to escape; only Syrians and Eritreans are not fingerprinted. All the others are identified and taken to the hosting centres where they can claim asylum and, according to Dublin III regulation, are not allowed to demand international protection in another EU member state.

References

Agamben, Giorgio. 1998. *Homo Sacer: Sovereign Power and Bare Life*. Stanford, CA: Stanford University Press.

Agnew, Jonathan, 1994, 'The Territorial Trap: The Geographical Assumptions of International Relations Theory'. *Review of International Political Economy* 1: 53–80.

Aita, Samir. 2013. 'Lo tsunami dei giovani: I paesi Arabi tra rivolta, migrazione e sfruttamento'. Accessed 18 March 2014. http://www.connessioniprecarie.org/2013/02/18/lo-tsunami-dei-giovani-i-paesi-arabi-tra-rivolta-migrazione-e-sfruttamento/

Amin, Samir, *The People's Spring: The Future of the Arab Revolution*. Cape Town: Pambazuka Press, 2012.

Ammor, Fouad M. 2012 'Good Democratic Governance in the Mediterranean'. In *Opportunities in the Emerging Mediterranean*, edited by Stephen Calleya and M. Wohlfeld, 120–32. Msida, Malta: Mediterranean Academy of Diplomatic Studies.

Amoore, Louise. 2013. *The Politics of Possibility: Risk and Security beyond Probability*. Durham, NC: Duke University Press.

Anderson, Benedict. 1991. *Imagined Communities: Reflections on the Origin and Spread of Nationalism*. London: Verso.

Anderson, Bridget, Nandita Sharma and Cinzia Wright. 2011. 'Editorial: Why No Borders?' *Refugee Studies* 26(2).

Andersson, Ruben. 2012. 'A Game of Risk: Boat Migration and the Business of Bordering Europe'. *Anthropology Today* 28(6): 7–11.

Andrijasevic, Rutvica, and William Walters. 2010. 'The International Organization for Migration and the International Government of Borders'. *Environment and Planning D: Society and Space* 28(6): 977–99.

Aradau, Claudia, and Rens Van Munster. 2007. 'Governing Terrorism through Risk: Taking Precautions, (Un)knowing the Future'. *European Journal of International Relations* 13(1): 89–115.

Aradau, Claudia, and Tobias Blanke. 2010. 'Governing Circulation: A Critique of the Biopolitics of Security'. In *Security and Global Governmentality: Globalization, Governance and the State*, edited by Miguel de Larrinaga and Marc G. Doucet. London: Routledge.

Asad, Talal. 2003. *Formations of the Secular: Christianity, Islam, Modernity*. Stanford, CA: Stanford University Press.

———. 2009. 'Free Speech, Blasphemy and Secular Criticism'. In *Is Critique Secular? Blasphemy, Injury, and Free Speech*, edited by Talal Asad, Wendy Brown, Judith Butler and Saba Mahmood, Townsend Papers in the Humanities, vol. 2, 20–63. Berkeley, CA: Doreen B. Townsend Center for the Humanities.

Athanasiaou, Athena, and Judith Butler. 2013. *Dispossession: The Performative in the Political*. Cambridge: Polity Press.

Azouray, Ariela. 2008. *The Civil Contract of Photography*. New York: Zone Books.

Badiou, Alain. 2013. *The Rebirth of History*. London: Verso.

Balibar, Étienne. 2002. *Politics and the Other Scene*. London: Verso.

———. 2004. *We the People of Europe? Reflections on Transnational Citizenship*. Princeton, NJ: Princeton University Press.

———. 2009. 'Europe as Borderland'. *Environment and Planning D: Society and Space* 27(2): 190–215.

———. 2010a. *Violence et civilité*. Paris: Galilée.

———. 2010b. 'At the Borders of Citizenship: A Democracy in Translation?' *European Journal of Social Theory*, 13(3): 315–22.

———. 2010c. *La proposition de l'egaliberté*. Paris: Actuel Marx.

———. 2013. 'On the Politics of Human Rights'. *Constellation s* 20: 8–26.

Balibar, Étienne, and Alain Brossat. 2011. 'Les voix des deux rives: A propos du Printemps Arab et des migrations'. *Outis* 1: 143–62.

Balibar Étienne, Sandro Mezzadra and Ranabbir Samaddar, eds. 2012. *The Borders of Justice*. Philadelphia: Temple University Press.

Barry Andrew, Thomas Osborne, Rose Nikolas, eds. 1996. *Foucault and Political Reason: Liberalism, Neoliberalism and Rationalities of Government*. Chicago: University of Chicago Press.

Basaran, Tugba. 2011. *Security, Law and Borders: At the Limits of Liberties*. Abingdon, UK: Routledge.

Basbous, Antoine. 2011. *Le tsunami Arabe*. Paris: Fayard.

Beneduce Roberto. 2008. 'Undocumented Bodies, Burned Identities: Refugees, Sans Papiers, Harraga. When Things Fall Apart'. *Social Sciences and Information* 47: 505–27.

Benjamin, Walter. 1969. *Illuminations*. New York: Schocken.

———. 1986. 'Critique of Violence'. In *Reflections*. New York: Schocken.

Benhabib, Sheyla. 2011. 'The Arab Spring: Religion, Revolution and the Public Sphere'. *Eurozine*. Accessed 20 March 2014. http://www.eurozine.com/pdf/2011-05-10-benhabib-en.pdf

Betts, Alexander. 2010a. 'Soft Law and the Protection of Vulnerable Migrants'. Accessed 18 March 2014. http://scholarship.law.georgetown.edu/imbr_2010/3

———. 2010b. 'Survival Migration: A New Protection Framework'. *Global Governance* 16, 361–82.

———, ed. 2011. *Global Migration Governance*. Oxford: Oxford University Press.

Bhabha, Hohmi. 1990. 'DissemiNation: Time, Narrative, and the Margins of the Modern Nation'. In *Nation and Narration*, 291–322. London: Routledge.

Bialasiewicz, Luiza, Paolo Giaccaria, Alun Jones and Claudio Minca. 2013. 'Re-scaling 'EU'rope: EU Macro-Regional Fantasies in the Mediterranean'. *European Urban and Regional Studies* 20: 59–76.

Biemann, Ursula, and Brian Holmes, eds. 2006. *The Maghreb Connection: Movements of Life across North Africa*. Barcelona: Actar.

Bigo, Didier. 2002. 'Security and Immigration: Towards a Critique of the Governmentality of Unease'. *Alternatives* 27(1): 63–92.

———. 2005. 'Global (in)Security: The Field of the Professionals of Unease Management and the Ban-Opticon'. *Traces* 4.

———. 2008. 'Globalized (in)Security: The Field and the Ban-Opticon'. In *Terror, Insecurity and Liberty: Illiberal Practices of Liberal Regimes after 9/11*, edited by Bigo Didier and Anastasia Tsoukala, 10–48. London: Routledge.

———. 2011. 'Freedom and Speed in Enlarged Borderzones'. In *The Contested Politics of Mobility: Borderzones and Irregularity*, edited by Vicki Squire, 31–50. London: Routledge.

Bigo, Didier, and Elspeth Guild, eds. 2005. 'Controlling Frontiers: Free Movement into and within Europe', edited by Didier Bigo and Elspeth Guild. Aldershot, UK: Ashgate.

Bigo, Didier, and Elspeth Guild. 2010. 'The Transformation of European Border Controls'. In *Extraterritorial Immigration Control: Legal Challenges*, edited by Bernhard Ryan and Valsamis Mitsilegas. Leiden: Martinus Nijhoff.

Bishara, Marwan. 2012. *The Invisible Arab: The Promise and Peril of the Arab Revolutions*. New York: Nation Books.

Black, Jeremy. 2002. *Maps and Politics*. Chicago: University of Chicago Press.

Blunt, Alison. 2007. 'Cultural Geographies of Migration: Mobility, Transnationality and Diaspora'. *Progress in Human Geography* 31(5): 684–94.

Bohmer Carol, and Amy Shuman. 2008. *Rejecting Refugees: Political Asylum in the 21st Century*. London: Routledge.
Bojadžijev Manuela, and Isabelle Saint-Saëns. 2009. 'Borders, Citizenship, War, Class: A Discussion with Étienne Balibar and Sandro Mezzadra'. Accessed 25 March 2014. www.academia.edu/3197478/Sandro_Mezzadra_and_Etienne_Balibar_Borders_Citizenship_War_Class._A_Dialogue_2006_
Bojadžijev, Manuela, and Serhat Karakayali. 2010. 'Recuperating the Sideshows of Capitalism: The Autonomy of Migration Today'. *e-flux journal* 17. Accessed 20 February 2014. http://worker01.eflux.com/pdf/article_154.pdf
Boswell, Christina. 2003. 'The "External Dimension" of EU Immigration and Asylum Policy'. *International Affairs* 79(3): 619–38.
Boubakri, Hassan. 2013. 'Revolution and International Migration in Tunisia'. *Migration Policy Center Report* 4.
Brenner, Neil. 2004. *New State Spaces: Urban Governance and the Rescaling of Statehood*. Oxford: Oxford University Press.
Brockling, Ulrich, Susanne Krasmann and Thomas Lemke, eds. 2011. *Governmentality: Current Issues and Future Challenges*. New York: Routledge.
Broeders, Dennis. 2007. 'The New Digital Borders of Europe: EUdatabases and the Surveillance of Irregular Migrants'. *International Sociology* 22(1): 71–92.
Brown, Wendy. 2005. '"The Most We Can Hope For . . ." Human Rights and the Politics of Fatalism'. *South Atlantic Quarterly* 103(2–3): 451–63.
Buck-Morss, Susan. 2009. *Hegel, Haiti, and Universal History*. Pittsburgh, PA: University of Pittsburgh Press.
Burchell Graham, Colin Gordon and Peter Miller, eds. 1991. *The Foucault Effect: Studies in Governmentality*. Chicago: University of Chicago Press.
Butler, Judith. 1997. *Excitable Speech: A Politics of the Performative*. New York: Routledge.
———. 2006a. *Precarious Life: The Powers of Mourning and Violence*. London: Verso.
———. 2006b. *Gender Trouble: Feminism and the Subversion of Identity*, Abingdon, UK: Routledge.
———. 2009. *Frames of War: When Is Life Grievable?* London: Verso.
Bygrave, Stephen, and Stephen Morton, eds. 2008. *Foucault in an Age of Terror: Essays on Biopolitics and the Defence of Society*. New York: Palgrave Macmillan.
Carrera, Sergio, Lehonard den Hertog and Joanna Parkin. 2012. 'EU Migration Policy in the Wake of the Arab Spring: What Prospects for EU–Southern Mediterranean Relations?' *MEDPRO Technical Report* 15. Accessed 26 March 2014. http://www.ceps.eu/book/eu-migration-policy-wake-arab-spring-what-prospects-eu-southern-mediterranean-relations
Casas-Cortes, Maria, Sebastian Cobarrubias and John Pickles. 2011. 'An Interview with Sandro Mezzadra'. *Environment and Planning D: Society and Space* 29: 584–98.
———. 2013. 'Re-bordering the Neighbourhood: Europe's Emerging Geographies of Non-accession Integration'. *European Urban and Regional Studies* 20(1): 37–58.
Cassarino, Jean-Pierre. 2012. 'Hiérarchie de priorités et système de réadmission dans les relations bilatérales de la Tunisie avec les états membres de l'Union Européenne'. In *Maghreb et Sciences Sociales*, 245–61. Paris: Institut de la recherche sur le Maghreb contemporain.
Cassarino, Jean-Pierre, and Sandra Lavenex. 2012. 'EU-Migration Governance in the Mediterranean Region: The Promise of (a Balanced) Partnership?' In *IEMed Mediterranean YearBook*, 284–88. Barcelona: IEMed.
Castles, Stephen, and Raul Delgado, eds. 2008. *Migration and Development: Perspectives from the South*. Geneva: IOM.
Ceriani, Paolo, et al. 2009. 'Report on the Situation on the Euro-Mediterranean Borders (from the Point of View of the Respect of Human Rights)'. Accessed 18 March 2014. http://www.libertysecurity.org/article2497.html
Chakrabarty, Dipesh. 2000a. *Provincializing Europe: Postcolonial Thought and Historical Difference*. Princeton, NJ: Princeton University Press.

———. 2000b. 'Radical Histories and Question of Enlightenment Rationalism: Some Critiques of Subaltern Studies'. In *Mapping Subaltern Studies and the Postcolonial*, edited by Vinayak Chaturvedi, 256–80. London: Verso.
Chaloff, Jonathan. 2007. 'Co-Development: A Myth or a Workable Recent Policy Approach?'. CESPI. www.cespi.it/SCM/CoDev-Myth.pdf
Chamayou, Gregoire. 2010. *Le cacce all'uomo*. Roma: IlManifesto.
Chandler, David. 2009. 'Critiquing Liberal Cosmopolitanism? The Limits of the Biopolitical Approach'. *International Political Sociology* 3(1): 53–70.
Chatterjee, Partha. 1993. *Nationalist Thought and the Colonial World: A Derivative Discourse*. Minneapolis: University of Minnesota Press.
———. 2004. *The Politics of the Governed: Reflections on Popular Politics in Most of the World*. New York: Columbia University Press.
———. 2005. 'Sovereign Violence and the Domain of the Political'. In *Sovereign Bodies: Citizens, Migrants, and States in the Postcolonial World*, edited by Thomas B. Hansen and Finn Stepputat, 82–102. Princeton, NJ: Princeton University Press.
———. 2012. *Lineages of Political Society: Studies in Postcolonial Democracy*. New York: Columbia University Press.
Chen, Kuan-Hsing. 2010. *Asia as Method: Toward Deimperialization*. Durham, NC: Duke University Press.
Chignola, Sandro, and Sandro Mezzadra. 2012. 'Fuori dalla politica: Laboratori globali della soggettività', *Filosofia Politica* 1: 65–81.
Cobarrubias, Sebastian. 2009. 'Mapping Machines: Activist Cartographies of the Border and Labor Lands of Europe'. PhD dissertation, Duke University.
Cobarrubias, Sebastian, and John Pickles. 2009. 'Spacing Movements: Mapping Practices, Global Justice and Social Activism'. In *Spatial Turn*, edited by Barney Warf and Santa Arias, 36–58. London: Routledge.
Colectivo Situaciones. 2003. 'On the Researcher-Militant', European Institute for Progressive Cultural Policies (EIPCP). Accessed 20 March 2014. http://eipcp.net/transversal/0406/colectivosituaciones/en
———. 2007. 'Something More on Research Militancy: Footnotes on Procedures and (In)Decisions'. In *Constituent Imagination: Militant Investigations, Collective Theorization*, edited by Stevphen Shukaitis and David Graeber, 73–93. Oakland, CA: AK Press.
Collier, Michael, and Aihwa Ong, ed. 2005. *Technology, Politics and Ethics as Anthropological Problems*. Malden, MA: Blackwell.
Conrad Joseph. 2007. *Heart of Darkness*. London: Penguin.
Cortes, Jose-Miguel, et al., eds. 2008. *Dissident Cartographies*. Barcelona: Actar.
Cotoi, Calin. 2011. 'Neoliberalism: A Foucaultian Perspective'. *International Review of Social Research* 1(2): 109–24.
Counter Cartographies Collective. 2012. 'Counter (Mapping) Actions: Mapping as Militant Research'. *Acme*, 11(3): 439–66.
Coutin, Barbara. 2011. 'Legal Exclusion and Dislocated Subjectivities: The Deportation of Salvadoran Youth from the United States'. In *The Contested Politics of Mobility: Borderzones and Irregularity*, edited by Vicki Squire, 169–83. London: Routledge.
Crampton, Jeremy, and John Krigyer. 2006. 'An Introduction to Critical Cartography'. *Acme* 4(1): 11–33.
Crampton Jeremy, and Stuart Elden, eds. 2007. *Space, Knowledge and Power: Foucault and Geography*. London: Ashgate.
Cuttitta Paolo. 2007. *Segnali di confine: Il controllo dell'immigrazione nel mondo Ffrontiera*. Milano: Mimesis Edizioni, Milano.
———. 2010. 'Readmission in the Relations between Italy and North African Mediterranean Countries'. In *Unbalanced Reciprocities: Cooperation on Readmission in the Euro-Mediterranean Area*, edited by Jean-Pierre Cassarino, Middle East Institute. Accessed 15 March 2014. http://www.statewatch.org/news/2010/sep/eu-unbalanced-reciprocities-middle-east-institute.pdf

———. 2012. *Lo spettacolo del confine: Lampedusa tra produzione e messa in scena della frontiera*. Milano: Mimesis Edizioni.
Dabashi, Hamid. 2012. *The Arab Spring: The End of Postcolonialism*. London: Zed Books.
Davidson, Arnold. 2004. *The Emergence of Sexuality: Historical Epistemology and the Formation of Concepts*. Cambridge, MA: Harvard University Press.
———. 2013. 'Des jeux linguistiques à l'epistemologie politique'. In *Le savoir sans fondements* by Aldo Giorgio Gargani. Paris: Vrin.
Dean, Mitchell. 1999. *Governmentality: Power and Rules in Modern Societies*. Thousand Oaks, CA: Sage.
De Genova, Nicholas. 2004. 'The Legal Production of Mexican/Migrant Illegality'. *Latino Studies* 2, 160–85.
———. 2005. *Working the Boundary: Race, Space, and Illegality in Mexican Chicago*. Durham, NC: Duke University Press, 2005.
———. 2010a. 'The Queer Politics of Migration: Reflections on 'Illegality' and Incorrigibility', *Studies in Social Justice* 4(2): 101–26.
———. 2010b. 'The Deportation Regime. Sovereignty, Space and the Freedom of Movement'. In *The Deportation Regime: Sovereignty, Space, and the Freedom of Movement*, edited by Nicholas De Genova and Natalie Peutz, 33–65. Durham, NC: Duke University Press.
———. 2011. 'Alien Powers. Deportable Labour and the Spectacle of Security'. In *The Contested Politics of Mobility: Borderzones and Irregularity*, edited by Vicki Squire, 91–113. London: Routledge.
———. 2013a. 'Spectacles of Migrant "Illegality": The Scene of Exclusion, the Obscene of Inclusion'. *Ethic and Racial Studies*. Accessed 20 March 2014. DOI:10.1080/01419870.2013.783710
———. 2013b. '"We Are of the Connections": Migration, Methodological Nationalism, and Militant Research', *Postcolonial Studies* 16(3): 250–58.
De Genova, Nicholas, and Natalie Peutz, eds. 2010. *The Deportation Regime: Sovereignty, Space, and the Freedom of Movement*. Durham, NC: Duke University Press.
De Haas, Hein. 2007a. 'Turning the Tide? Why Development Will Not Stop Migration'. *Development and Change* 38(5): 821–44.
———. 2007b. 'North African Migration Systems: Evolution, Transformations and Development Linkages'. International Migration Institute Working Paper no. 6.
———. 2008. 'The Myth of Invasion: The Inconvenient Realities of Migration from Africa to the European Union'. *Third World Quarterly* 29(7): 1305–22.
Deleuze, Gilles, and Felix Guattari. 1975. *Kafka: Pour une Littérature Mineure*. Paris: Editions de Minuit.
———. 2007. *A Thousand Plateaux: Capitalism and Schizophrenia*. New York: Continuum.
Dembour, Marie-Benedict, and Tobias Kelly Tobias. 2011. *Are Human Rights for Migrants? Critical Reflections on the Status of Irregular Migrants in Europe and the United States*. Abingdon, UK: Routledge.
Derrida, Jacques. 2005. *Writing and Difference*. Chicago: University of Chicago Press.
———. 2006. *Specters of Marx: The State of the Debt, the Work of Mourning and the New International*. Abingdon, UK: Routledge.
Douzinas, Costas. 2007. *Human Rights and Empire: The Political Philosophy of Cosmopolitanism*. Abingdon, UK: Routledge.
Dillon, Michael. 2007. 'Governing through Contingency: The Security of Biopolitical Governance', *Political Geography* 26: 41–47.
Dillon, Michael, and Andrew Neal, eds. 2008. *Foucault on Politics, Security and War*. London: Palgrave MacMillan.
Elden, Stuart. 2007. 'Governmentality, Calculation, Territory', *Environment and Planning D: Society and Space* 25(3): 562–80.
Escobar, Arturo. 1996. *Encountering Development: The Making and Unmaking of the Third World*. Princeton, NJ: Princeton University Press.

Eyadat, Zaid. 2012. 'The Arab Revolutions of 2011: Revolutions of Dignity'. In *Change and Opportunities in the Emerging Mediterranean*, edited by Stephen Calleya and Monika Wohlfeld, 3–19. Msida, Malta: Mediterranean Academy of Diplomatic Studies.

Faist, Thomas. 2008. 'Migrants as Transnational Development Agents: An Inquiry into the Newest Round of the Migration-Development Nexus'. *Population, Place, Space* 14(1): 21–42.

Faist, Thomas, Margit Fauser and Peter Kivisto. 2010. *The Migration-Development Nexus: A Transnational Perspective*. New York: Palgrave Macmillan.

Fanon, Frantz. 1994a. *Towards the African Revolution*. New York: Grove Press.

———. 1994b. *A Dying Colonialism*. New York: Grove Press.

———. 2007. *The Wretched of the Earth*. New York: Grove Press.

———. 2011. *Decolonizzare la Follia: Scritti sulla Psichiatria Coloniale*. Verona: Ombre Corte.

Fargue, Philippe, and Christine Fandrich. 2012. *Migration after the Arab Spring*. Migration Policy Centre Research Report 9. Accessed 19 March 2014. http://cadmus.eui.eu/bitstream/handle/1814/23504/MPC-RR-2012-09.pdf?sequence=1

Fassin, Didier. 2011a. *Humanitarian Reason: A Moral History of the Present*. Berkeley: University of California Press.

———. 2011b. 'Policing Borders, Producing Boundaries: The Governmentality of Immigration in Dark Times'. *Annual Review of Anthropology* 40: 213–26.

———. 2013. 'The Precarious Truth of Asylum'. *Public Culture* 25(1): 39–63.

Fassin, Didier, and Ester d' Halluin. 2005. 'The Truth from the Body. Medical Certificates as Ultimate Evidence for Asylum Seekers'. *American Anthropology* 107(4): 597–608.

Fassin, Dider, and Richard Rechtman. 2009. *The Empire of Trauma: An Inquiry into the Condition of Victimhood*. Princeton, NJ: Princeton University Press.

Fassin, Didier, and Pandolfi, Mariella. 2010. *Contemporary States of Emergency: The Politics of Military and Humanitarian Intervention*. London: Zone Books.

Feher, Michel, ed. 2007. *Nongovernmental Politics*. New York: Zone Books.

Feldman, Gregory. 2011. *The Migration Apparatus: Security, Labor, and Policymaking in the European Union*. Stanford, CA: Stanford University Press.

Fekete, Liza. 2005. 'The Deportation Machine: Europe, Asylum and Human Rights'. *Race and Class* 47(1): 64–91.

Filali-Ansary, Abddou. 2012. 'The Languages of the Arab Revolutions', *Journal of Democracy* 23(2): 5–18.

Fischer-Lescano, Andreas, Tillmann Lohr and Timo Tohidipur. 2009. 'Border Controls at Sea: Requirements under International Human Rights and Refugee Law'. *International Journal of Refugee Law* 21(2): 256–96.

Foucault, Michel. 1980a. 'Power and Strategies'. In *Power/Knowledge: Selected Interviews and Other Writings, 1972–1977*, edited by Gordon Colin, 134–45. London: Harvester Press.

———. 1980b. 'The Confession of the Flesh'. In *Power/Knowledge: Selected Interviews and Other Writings, 1972–1977*, 194–228. London: Harvester Press.

———. 1982. 'The Subject and Power'. *Critical Inquiry* 8(4): 777–95.

———. 1984a. 'Nietzsche, Genealogy, History'. In *The Foucault Reader*, edited by Paul Rabinow, 76–100. New York: Pantheon Books.

———. 1984b. 'What Is Enlightenment?' In *The Foucault Reader*, edited by Paul Rabinow, 32–50. New York: Pantheon Books.

———. 1988. 'Technologies of the Self'. In *Technologies of the Self: A Seminar with Michel Foucault*, edited by Luther H. Martin, Huck Gutman and Patrick H. Hutton, 16–49. Amherst: University of Massachusetts Press.

———. 1994a. 'Lives of Infamous Men'. In *Power, Volume 3 of The Essential Works of Michel Foucault 1954–1984*, edited by James D. Faubion, 57–175. London: Penguin.

———. 1994b. 'Omnes et Singulatim': Towards a Critique of Political Reason'. In *Power, Volume 3 of The Essential Works of Michel Foucault 1954–1984*, edited by James D. Faubion, 298–325. London: Penguin.

———. 1994c. 'Space, Knowledge and Power'. In *Power, Volume 3 of The Essential Works of Michel Foucault 1954–1984*, edited by James D. Faubion, 346–64. London: Penguin.
———. 1994d. 'Useless to Revolt?' In *Power, Volume 3 of The Essential Works of Michel Foucault 1954–1984*, edited by James D. Faubion, 449–53. London: Penguin Books.
———. 1994e. 'The Masked Philosopher'. In *Ethics: Subjectivity and Truth, Volume 2 of The Essential Works of Michel Foucault 1954–1984*, edited by Paul Rabinow, 321–28. London: Penguin Books.
———. 1994f. 'The Social Triumph of the Sexual Will: A Conversation with Michel Foucault'. In *Ethics: Subjectivity and Truth, Volume 2 of the Essential Works of Michel Foucault 1954–1984*, edited by Paul Rabinow, 167–72. New York: New Press.
———. 1994g. 'So Is It Important to Think?' In *Power, Volume 3 of the Essential Works of Michel Foucault 1954–1984*, edited by James D. Faubion, 454–58. New York: New Press.
———. 1995. *Discipline and Punish: The Birth of the Prison*. New York: Vintage Books.
———. 1996a. 'An Ethics of the Concern of Self'. *Foucault Live: Collected Interviews*, edited by Sylvère Lotringer, 432–49. New York: Semiotexte.
———. 1996b. 'Friendship as a Way of Life'. In *Foucault Live: Collected Interviews, 1961–1984*, edited by Lotringer Sylvère, 308–12. New York: Semiotext(e).
———. 1996c. 'The Social Extension of the Norm'. In *Foucault Live: Collected Interviews , 1961–1984*, edited by Lotringer Sylvère, 196–99. New York: Semiotext(e).
———. 1997. 'What Is Critique?' In *The Politics of Truth*, edited by Lotringer Sylvère and John Rajchman, 41–82. New York: Semiotext(e).
———. 1998. *The History of Sexuality: The Will to Knowledge*. London: Penguin Books.
———. 2001a. 'Foucault reponde à Sartre'. In *Dits et écrits I*, 690–96. Paris: Gallimard.
———. 2001b. 'Réponse à une question'. In *Dits et écrits I*, 700–723. Paris: Gallimard.
———. 2001c. 'Sur les prisons'. In *Dits et écrits I*, 1044–60. Paris: Gallimard.
———. 2001d. 'Le monde est un grand asile'. In *Dits et écrits I*, 1298–1309. Paris: Gallimard.
———. 2001e. 'Défi à l'opposition'. In *Dits et écrits II*, 704–6. Paris: Gallimard.
———. 2001f. 'La philosophie analytique de la politique'. In *Dits et écrits II*, 534–51. Paris: Gallimard.
———. 2001g. 'Le chef mythique de la révolte en Iran'. In *Dits et écrits II*, 713–15. Paris: Gallimard.
———. 2001h. 'Précisations sur le pouvoir'. In *Dits et écrits II*, 625–35. Paris: Gallimard.
———. 2001i. 'Table ronde du 20 Mai 1978'. In *Dits et écrits II*, 839–53. Paris: Gallimard.
———. 2001l. 'Inutile de se soulever?' In *Dits et écrits II*, 790–94. Paris: Gallimard.
———. 2003a. *Society Must Be Defended: Lectures at the Collège de France, 1975–76*. New York: Picador.
———. 2003b. *Abnormal: Lectures at the College de France, 1974–1975*. London: Verso.
———. 2006a. *Psychiatric Power: Lectures at the College de France, 1973–1974*. New York: Picador.
———. 2006b. *History of Madness*. London: Routledge.
———. 2009. *Security, Territory, Population: Lectures at the Collège de France, 1977–1978*. New York: Picador/Palgrave Macmillan.
———. 2010. *The Birth of Biopolitics: Lectures at the Collège de France, 1978–1979*. New York: Picador.
———. 2012a. *Mal faire, dire vrai: Fonction de l'aveu en justice*. Louvain: Presses Universitaires de Louvain.
———. 2012b. *Sull'Origine dell'Ermeneutica del Sé*. Napoli: Cronopio.
———. 2013a. *La société punitive: Cours au Collège de France , 1972–1973*. Paris: Gallimard.
———. 2013b. 'Subjectivité et vérité'. In *Sur l'origine de l' hermeneutique de soi*, 107–21. Paris: Vrin.
———. 2014. *Subjectivité et vérité: Cours au Collège de France, 1980–1981*. Paris: Gallimard.
Galli, Carlo. 2001. *Spazi Politici: L'età moderna e l'Età Globale*. Bologna: Il Mulino.

Gammeltoft-Hansen, Thomas. 2008. 'The Refugee, the Sovereign and the Sea: EU Interdiction Policies in The Mediterranean'. DIIS Working Paper no. 2008/6.
Garelli, Glenda. 2013. 'Schengen Intermittences: The On/Off Circuit of Free Circulation'. In *Spaces in Migration: Postcards of a Revolution*, edited by Glenda Garelli, Federica Sossi and Martina Tazzioli. London: Pavement Books.
Garelli, Glenda, Martina Tazzioli. 2013. 'Migration Discipline Hijacked: Distances and Interruptions of a Research Militancy'. *Postcolonial Studies* 16(3): 299–308.
Geiger, Martin. 2013. 'The Transformation of Migration Politics: From Migration Control to Disciplining Mobility'. In *Disciplining the Transnational Mobility of People*, edited by Martin Geiger and Antoine Pécoud, 15–40. New York: Palgrave Macmillan.
Geiger, Martin, and Antoine Pécoud, eds., 2010. *The Politics of International Migration Management*. London: Palgrave Macmillan.
Georgi, Fabian. 2010. 'For the Benefit of Some: The International Organization for Migration and Its Global Migration Management'. In *The Politics of International Migration Management*, edited by Martin Geiger and Antoine Pécoud, 45–62. New York: Palgrave Macmillan.
Ghosh, Bimal. 2010. 'Managing Migration: Towards the Missing Regime?' In *Migration without Borders: Essays on the Free Movement of People*, edited by Antoine Pécoud and Paul de Guchteneire, 97–118. New York: Berghahn.
———. 2012. *The Global Economic Crisis and the Future of Migration: Issues and Prospects*. New York: Palgrave Macmillan.
Giddens, Anthony. 1973. *The Class Structure of the Advanced Societies*. London: Hutchinson.
Gilroy, Paul. 1993. *The Black Atlantic: Modernity and Double Consciousness*. London: Verso.
Goswami, Manu. 2004. *Producing India: From Colonial Economy to National Space*. Chicago: University of Chicago Press.
Grant, Stephanie. 2011a. 'Recording and Identifying European Frontier Deaths'. *European Journal of Migration and Law* 13: 135–56.
———. 2011b. 'Migrant Deaths at Sea: The Challenge to Record and Identify'. *New Europe*. Accessed 18 March 2014. http://www.neurope.eu/article/migrant-deaths-sea-challenge-record-and-identify
Gregory, Derek. 1994. *Geographical Imaginations*. Oxford: Blackwell.
Good, Anthony. 2004. 'Expert Evidence in Asylum and Human Rights Appeal: An Expert's View'. *International Journal of Refugee Law* 16(3): 359–80.
Guha, Ranajit. 1988. 'The Prose of Counterinsurgency'. In *Selected Subaltern Studies*, ed. Ranajit Guha and Gayatri C. Spivak, 45–88. Oxford: Oxford University Press.
———. 2003. *History at the Limit of World-History*. New York: Columbia University Press.
Haar Jens H., and William Walters. 2005. *Governing Europe: Discourse, Governmentality and European Integration*. London: Routledge.
Hacking, Ian. 2004a. *Language, Truth and Reason in Historical Ontology*. Cambridge, MA: Harvard University Press.
———. 2004b. 'Making Up People'. In *Historical Ontology*. Cambridge, MA: Harvard University Press.
Hailbronner, Kay. 1993. 'The Concept of 'Safe Country' and Expeditious Asylum Procedures: A Western European Perspective'. *International Journal of Refugee Law* 5(1): 31–65.
Hansen, Thomas B., and Finn Stepputat, eds. 2005. *Sovereign Bodies: Citizens, Migrants, and States in the Postcolonial World*. Princeton, NJ: Princeton University Press.
Harley, Brian.1989. 'Deconstructing the Map'. *Cartographica* 26(2): 1–20.
———. 2001. *The New Nature of Maps: Essays in the History of Cartography*. Baltimore: Johns Hopkins University Press.

Heller, Charles, and Pezzani Lorenzo. 2014. 'Liquid Traces: Investigating the Deaths of Migrants at the EU's Maritime Frontier'. In *Forensis: The Architecture of Public Truth*, edited by Eyal Weizman, New York: Stenberg Press.

Hess, Sabine. 2008. 'Migration and Development: A Governmental Twist of the EU Migration Management Policy'. Paper presented at Sussex Centre for Migration Research.

———. 2012. 'De-naturalising Transit Migration: Theory and Methods of an Ethnographic Regime Analysis'. *Population, Space and Place* 18(4): 428–40.

Hess, Sabine, Serhat Karakayali and Vassilis Tsianos. 2009. 'Transnational Migration Theory and Method of an Ethnographic Analysis of Border Regimes'. Working Paper, Sussex Centre for Migration Research. www.sussex.ac.uk/webteam/gateway/file.php?name=mwp55.pdf&site=252

Hindess, Barry. 1997. 'Politics and Governmentality'. *Economy and Society* 26(2): 257–72.

Honig, Bonnie. 2001. *Democracy and the Foreigner*. Princeton, NJ: Princeton University Press.

Huysmans, Jef. 2006. *The Politics of Insecurity: Fear, Migration and Asylum in the EU*. Abingdon, UK: Routledge.

Hyndman, Jennifer. 2000. *Managing Displacement: Refugees and the Politics of Humanitarianism*. Minneapolis: University of Minnesota Press.

Huggan, Graham. 1994. *Territorial Disputes: Maps and Mapping Strategies in Contemporary Canadian and Australian Fiction*. Toronto: University of Toronto Press.

Korzybsky, Alfred. 1933. 'A Non-Aristotelian System and Its Necessity for Rigour in Mathematics and Physics'. In *Science and Sanity*, Supplement III, 747–61. Englewood, NJ: Institute of General Semantics.

Kratochwil, Friedrich. 2011. 'Of Maps, Law, and Politics: An Inquiry into the Changing Meaning of Territoriality'. DIIS Working Paper no. 3. Accessed 25 March 2014. http://www.econstor.eu/bitstream/10419/44626/1/647472775.pdf

Inda, Jonathan X., ed. 2005. *Anthropologies of Modernity: Foucault, Governmentality, and Life Politics*. Malden, MA: Blackwell.

———. 2006. *Targeting Immigrants: Government, Technology, and Ethics*. Oxford: Blackwell.

Irrera, Orazio, and Rada Ivekovic. 2013. Introduction. *Outis* 4: 1–13.

Isin, Engin. 2002. *Being Political: Genealogies of Citizenship*. Minneapolis: University of Minnesota Press.

———. 2008. 'Theorizing Acts of Citizenship'. In *Acts of Citizenship*, edited by Engin Isin and Greg M. Nielsen, 15–43. London: Palgrave Macmillan.

———. 2009. 'Citizenship in Flux: The Figure of the Activist Citizen'. *Subjectivity* 29: 367–88.

———. 2012. *Citizens without Frontiers*. London: Continuum.

Jabri, Vivienne. 2013. *The Postcolonial Subject: Claiming Politics/Governing Others in Late Modernity*. New York: Routledge.

Jacob, Christian. 2006. *The Sovereign Map: Theorethical Approaches in Cartography throughout History*. Chicago: University of Chicago Press.

Jeandesboz, Julien. 2011. 'Beyond the Tartar Steppe: EUROSUR and the Ethics of European Border Control Practices'. In *A Threat against Europe? Security, Migration and Integration*, edited by Peter Burgess and Serge Gutwirth, 111–32. Brussels: VUB Press.

Jones, Colin, and Roy Porter, eds. 1994. *Reassessing Foucault: Power, Medicine and the Body*. New York: Taylor and Francis.

Joseph, Jonathan. 2010. 'The Limits of Governmentality'. *European Journal of International Relations* 16(2): 223–46.

Kanna, Ahmed. 2011. 'The Arab World's Forgotten Rebellions: Foreign Workers and Biopolitics in the Gulf'. *Samar: The South Asian Magazine for Action and Reflection*. Accessed 20 March 2014. http://samarmagazine.org/archive/articles/357

Karakayali, Serhat, and Enrica Rigo. 2010. 'Mapping the European Space of Circulation'. In *The Deportation Regime: Sovereignty, Space, and the Freedom of Movement*, edited by Nicholas De Genova and Nathalie Peutz, 123–44. Durham, NC: Duke University Press.

Karakayali, Serhat, and Vassilis Tsianos. 2010. 'Transnational Migration and the Emergence of the European Border Regime: An Ethnographic Analysis'. *European Journal of Social Theory* 13(3): 373–87.

Kasparek, Bernd. 2010. 'Borders and Populations in Flux: Frontex's Place in the European Union's Migration Management'. In *The Politics of International Migration Management*, edited by Martin Geiger and Antoine Pécoud, 119–40. London: Palgrave Macmillan.

Kasparek, Bernd, and Fabian Wagner. 2012. 'Local Border Regimes or a Homogeneous External Border? The Case of the European Union's Border Agency Frontex'. In *The New Politics of International Mobility Migration Management and its Discontents*, edited by Martin Geiger and Antoine Pécoud, 173–92. Onasbruck, Germany: IMIS-Beitrage.

Kiersey, Nicholas. 2009. 'Neoliberal Political Economy and the Subjectivity of Crisis: Why Governmentality Is Not Hollow'. *Global Society* 23(4): 383–86.

King, George. 1996. *Mapping Reality: An Exploration of Cultural Cartographies*. New York: St. Martin's Press.

Koselleck, Renhart. 2012. *Crisi: Per un Lessico della Modernità*. Verona: Ombre Corte.

Kurz, Joshua J. 2011. '(Dis)locating Control: Transmigration, Precarity and the Governmentality of Control'. *Behemoth, a Journal on Civilisation*, 5(1): 30–51.

Kuster, Brigitte, and Vassilis Tsianos. 2013. 'Erase Them! Eurodac and Digital Deportability'. European Institute for Progressive Cultural Policies (EIPCP). Accessed 25 March 2014. http://eipcp.net/transversal/0313/kuster-tsianos/en

Kuster, Brigitte, Vassilis Tsianos, et al. 2012. Thematic Report 'Border Crossings': Transnational Digital Networks, Migration and Gender. MIG@NET. www.mignetproject.eu/wp-content/uploads/2012/10/MIGNET_Deliverable_6_Thematic_report_Border_crossings.pdf

Laclau, Ernesto. 2005. *On Populist Reason*. London: Verso.

———. 2007. *Emancipation(s)*. London: Verso.

Lauterpacht, Elihu, and Daniel Bethlehem. 2003. 'The Scope and Content of the Principle of Non-Refoulement'. Accessed 25 March 2014. http://www.refworld.org/cgi-bin/texis/vtx/rwmain?docid=470a33af0.

Lavenex, Sandra. 2008. 'A Governance Perspective on the European Neighbourhood Policy: Integration Beyond Conditionality?' *Journal of European Public Policy* 15(6): 938–55.

Lecadet, Clara. 2013. 'From Migrant Destitution to Self-organization into Transitory National Communities: The Revival of Citizenship in Post-Deportation Experience in Mali'. In *The Social, Political and Historical Contours of Deportation*, edited by Bridget Anderson, Matthew J. Gibney and Emanuela Paoletti, 143–58. New York: Springer.

Le Cour Grandmaison, Olivier. 2005. *Coloniser, exterminer: Sur la guerre et l'état colonial*. Paris: Fayard.

Lemke, Thomas. 2012. *Foucault, Governmentality, and Critique*. London: Paradigm.

———. 2013. 'Foucault, Politics and Failure'. In *Foucault, Biopolitics, and Governmentality*, edited by Jakob Nilsson and Svenolov Wallestein, 35–52. Huddinge, Sweden: Södertörns Högskola.

Leonard, Sarah. 2011. FRONTEX and the Securitization of Migrants through Practices. Accessed19 March 2014. European University Institute. http://www.eui.eu/Documents/RSCAS/Research/MWG/201011/SLeonardspaper.pdf

Liguori, Alessandro. 2008. *Le garanzie procedurali avverso l'espulsione degli immigrati in Europa*. Napoli: Editoriale Scientifica.

Likibi, Ramoulad. 2010. *L'Union Africaine face à la problématique migratoire*. Paris: Harmattan.

Lippert, Randy. 1999. 'Relevance of Governmentality to Understanding the International Refugee Regime'. *Alternatives* 24(3): 295–328.

Macherey, Pierre. 2013. *Il soggetto produttivo: Da Foucault a Marx*. Verona: Ombre Corte.

Lutterbeck, Derek. 2006. 'Policing Migration in the Mediterranean'. *Mediterranean Politics* 11(1): 59–82.

———. 2008. 'Coping with Europe's Boat People: Trends and Policy Dilemmas in Controlling the EU's Mediterranean Borders'. ISPI Policy Brief no. 76. http://www.ispionline.it/it/documents/PB_76_2008.pdf

Majidib Nassim, and Liza Schuster. 2013. 'What Happens Post-Deportation? The Experience of Deported Afghans', *Migration Studies* 1(2): 221–40.

Mahmood, Saba. 2005. *Politics of Piety: The Islamic Revival and the Feminist Subject*. Princeton, NJ: Princeton University Press.

Malkki, Lisa. 1992. 'National Geographic: The Rooting of People and the Territorialization of National Identity among Scholars and Refugees'. *Cultural Anthropology* 7(1): 24–44.

Malo, Marta. 2007. *Nociones comunes: Experiencias y ensayos entre investigaci n y militancia*. Madrid: Traficantes de Sueños.

Martins, Albrow. 1974. 'Time and Theory in Sociology'. In *Approaches to Sociology*, edited by Rex John, 246–94. London: Routledge & Kegan Paul.

Marx, Karl. 1993. *Grundrisse: Foundations of the Critique of Political Economy*. London: Penguin.

Marzouki, Nadia. 2011. 'From People to Citizens in Tunisia'. *Middle East Research and Information Project* 259. Accessed 20 March 2014. http://www.merip.org/mer/mer259/people-citizens-tunisia

Massad, Joseph. 2014. 'Love, Fear and the Arab Spring'. *Public Culture* 26: 127–52.

Massarelli, Fulvio. 2012. *La collera della Casbah*. Bologna: Agenzia X.

Mazzeo, Antonio. 2011. 'La partnership Italia-Tunisia contro i migranti'. Accessed 25 March 2014. http://www.storiemigranti.org/spip.php?rubrique125

Mbembe, Achille. 2003. 'Necropolitics'. *Public Culture* 15(1): 11–40.

McGrath, John. 2004. *Loving Big Brother: Performance, Privacy and Surveillance Space*. London: Routledge.

McKeown, Adam. 2008. *Melancholy Order: Asian Migration and the Globalization of Borders*. New York: Columbia University Press.

McLagan, Meg, and Yates McKee, eds. 2012. *Sensible Politics: The Visual Culture of Nongovernmental Activism*. New York: Zone Books.

McNevin, Angela. 2006. 'Political Belonging in a Neoliberal Era: The Struggle of the Sans-Papiers'. *Citizenship Studies* 10(2): 135–51.

———. 2011. Contesting Citizenship: Irregular Migrants and New Frontiers of the Political. New York: Columbia University Press.

Mezzadra, Sandro. 2006. *Diritto di fuga: Migrazioni, cittadinanza, globalizzazione*. Verona: OmbreCorte.

———. 2007. 'Living in Transition'. *Transversal* 11(7). Accessed 20 March 2014. http://translate.eipcp.net/transversal/1107

———. 2008. *La condizione postcoloniale: Storia e politica nel mondo globale*. Verona: Ombre Corte.

———. 2011. 'Le avventure mediterraneee della libertà'. In *Libeccio d'oltremare: Il vento delle rivoluzioni del Nord Africa si estende all'Occidente*, edited by Pirri Ambra. Roma: Ediesse Edizioni.

———. 2011b. 'En voyage: Michel Foucault et la critique post-coloniale'. In *Cahier Foucault*, edited by Philippe Artières et al., 352–57. Paris: Herne.

———. 2011c. 'The Gaze of Autonomy. Capitalism, Migration, and Social Struggles' Uninomade 2.0. Accessed 13 January 2014. http://www.uninomade.org/the-gaze-of-autonomy-capitalism-migration-and-social-struggles/

———. 2011d. 'How Many Histories of Labour? Towards a Theory of Postcolonial Capitalism'. *Postcolonial Studies* 14(2): 151–70.

———. 2014. *Nei cantieri marxiani*. Roma: IlManifesto.

Mezzadra, Sandro, and Brett Neilson. 2012. 'Borderscapes of Differential Inclusion: Subjectivity and Struggles on the Threshold of Justice's Excess'. In *The Borders of Justice,* edited by Étienne Balibar, Sandro Mezzadra and Ranabbir Samaddar, 181–204. Philadelphia: Temple University Press.
———. 2013. *Border as Method, or, The Multiplication of Labor*. Durham, NC: Duke University Press.
Mezzadra, Sandro, and Maurizio Ricciardi. 2013. *Movimenti indisciplinati: Migrazioni, migranti e discipline scientifiche*. Verona: Ombre Corte.
Mignolo, Walters. 2009. 'Epistemic Disobedience, Independent Thoughts, Decolonial Freedom'. *Theory, Culture & Society* 26(7–8): 1–23.
Mirzoeff, Nicholas. 2011. *The Right to Look: A Counterhistory of Visuality*. Durham, NC: Duke University Press.
Mitropoulos, Angela. 2007. 'Autonomy, Recognition, Movement'. In *Constituent Imagination: Militant Investigations, Collective Theorization*, edited by Stevphen Shukaitis, David Graeber, and Erika Biddle, 127–36. Oakland, CA: AK Press.
Mitropoulos, Angela, and Brett Neilson. 2006. 'Exceptional Times, Non-governmental Spacings, and Impolitical Movements'. *Vacarme* 34.
Mitsilegas, Valsamis, and Bernhard Ryan. 2010. *Extraterritorial Immigration Control: Legal Challenges*. Leiden: Martinus Nijhoff.
Mohanty, Chandra T. 2003. *Feminism without Borders: Decolonizing Theory, Practicing Solidarity*. Durham, NC: Duke University Press.
Mouffe, Chantal. 2005. *On the Political*. London: Routledge.
Moulier-Boutang, Yann. 2002. Dalla schiavitú al lavoro salariato. Roma: ManifestoLibri.
Murray Li, Tania. 2007. *The Will to Improve: Governmentality, Development, and the Practice of Politics*. Durham, NC: Duke University Press.
Nail, Thomas. 2013. 'The Crossroads of Power: Michel Foucault and the US/Mexico Border Wall'. *Foucault Studies* 15: 110–28.
Neal, Andrew W. 2009. 'Securitization and Risk at the EU Border: The Origins of FRONTEX'. *Journal of Common Market Studies* 47(2): 333–56.
Neilson, Brett, and Ned Rossiter. 2008. 'Precarity as a Political Concept'. *Open* 17, *A Precarious Existence*. Accessed 19 March 2014. http://www.skor.nl/_files/Files/OPEN17_P48-61(1).pdf
Neocleous, Mark. 2003. 'Off the Map: On Violence and Cartography'. *European Journal of Social Theory* 6(4): 409–20.
Noiriel, Gerard. 1991. *La tyrannie du national: Le droit d'asile en Europe , 1793–1993*. Paris: Calmann-Lévy.
Noiriel, Gerard. 1996. *The French Melting Pot: Immigration, Citizenship, and National Identity*. Minneapolis: University of Minnesota Press.
Nyers, Peter. 2003. 'Abject Cosmopolitanism: The Politics of Protection in the Anti-Deportation Movement'. *Third World Quarterly* 24(6): 1069–93.
———. 2006. 'The Accidental Citizen: Acts of Sovereignty and (Un)Making Citizenship'. *Economy and Society* 35(1): 22–41.
———. 2011. 'Forms of Irregular Citizenship'. In *The Contested Politics of Mobility: Borderzones and Irregularity*, edited by Vicki Squire. London: Routledge.
Nyers, Peter, and Kim Rygiel, eds. 2012. *Citizenship, Migrant Agency and the Politics of Movement*. London: Routledge.
Olsson, Gunnar. 2007. *Abysmal: A Critique of Cartographic Reason*. Chicago: University of Chicago Press.
O'Malley, Pat. 1996. 'Indigenous Governance'. *Economy & Society* 25: 310–26.
Ong, Aihwa. 2003. *Buddha Is Hiding: Refugees, Citizenship, the New America*. Berkeley: University of California Press.
———. 2006. *Neoliberalism as Exception: Mutations in Citizenship and Sovereignty*. Durham, NC: Duke University Press,
Paggi, Leonardo, ed. 2014. *Le rivoluzioni Arabe e le repliche della storia*. Verona: Ombre Corte.

Painter, Joe. 2008. 'Cartographic Anxiety and the Search for Regionality'. *Environment and Planning A* 40(2): 342–61.

Papadopoulos, Dimitris, Niamh Stephenson and Vassilis Tsianos. 2008. *Escape Routes: Control and Subversion in the 21st Century*. London: Pluto Press.

Papadopouls, Dimitris. 2010. 'Insurgent Posthumanism'. *Ephimera Journal*. Accessed 29 March 2014. http://www.ephemerajournal.org/sites/default/files/10-2papadopoulos_0.pdf

Papadopoulos, Dimitris, and Vassilis Tsianos. 2012. 'The Autonomy of Migration: The Animals of Undocumented Mobility'. European Institute for Progressive Cultural Policies (EIPCP). Accessed 28 March 2014. http://translate.eipcp.net/strands/02/papadopoulostsianos-strands01en

———. 2013. 'After Citizenship: Autonomy of Migration, Organisational Ontology and Mobile Commons'. *Citizenship Studies* 17(2): 178–96.

Papastergiadis, Nikos. 2000. *The Turbulence of Migration: Globalization, Deterritorialization and Hybridity*. Cambridge: Polity Press.

Pastore, Ferruccio. 2007. 'Europe, Migration and Development: Critical Remarks on an Emerging Policy Field'. CESPI. Accessed 20 March 2014. http://www.cespi.it/PDF/Pastore-MigrationandDevelopment.pdf

Patton, Paul. 1998. 'Foucault's Subject of Power'. In *The Later Foucault: Politics and Philosophy*, edited by Jeremy Moss, 64–77. New York: Sage Publications.

Peluso, Nancy. 1995. 'Whose Woods Are These? Counter-Mapping Forest Territories in Kalimantan, Indonesia'. *Antipode* 27(4): 383–406.

Peraldi, Michel. 2008. 'La condition migrant'. *La pensée de midi* 4(26): 81–94.

Perrin, Delphine. 2008. 'La circulation des personnes au Maghreb'. Cadmus, EUI Research Repository. http://cadmus.eui.eu/handle/1814/10090

Peters, Joel, eds. 2012. *The European Union and the Arab Spring: Promoting Democracy and Human Rights in the Middle East*. Lanham, MD: Lexington Books.

Peutz, Natalie. 2006. 'Embarking on an Anthropology of Removal'. *Current Anthropology* 47(2): 217–41.

———. 2010. 'Criminal Alien Deportees in Somaliland: An Ethnography of Removal'. In *The Deportation Regime: Sovereignty, Space, and the Freedom of Movement*, edited by Nicholas De Genova and Natalie Peutz, 371–412. Durham, NC: Duke University Press.

Pickering, Sharon, and Leanne Weber, eds. 2006. *Borders, Mobility and Technology of Control*. New York: Springer.

Pickles John. 2004. *A History of Spaces: Cartographic Reason, Mapping and the Geo-Coded World*. London: Routledge.

Piper, Nicola. 2009. 'The Complex Interconnections of the Migration–Development Nexus: A Social Perspective'. *Population, Space, Place* 15(2): 93–101.

Plascencia, Luis F. B. 2009. 'The 'Undocumented' Mexican Migrant Question: Re-examining the Framing of Law and Illegalization in the United States'. *Urban Anthropology* 38(2–4): 375–434.

Pollack, Kenneth, et al. 2011. *The Arab Awakening: America and the Transformation of the Middle East*. New York: Brookings Institution.

Potte-Bonneville, Matthieu. 2004. 'Politique des usages'. *Vacarme* 29.

Proctor, John. 1998. 'Ethics in Geography: Giving Moral Form to the Geographical Imagination'. *Area* 30(1): 8–18.

Pugh, Michael. 2001. 'Mediterranean Boat People: A Case for Co-operation?' *Mediterranean Politics* 6(1): 1–20.

Pugh, Michael. 2004. 'Drowning Not Waving: Boat People and Humanitarianism at Sea'. *Journal of Refugee Studies* 17: 50–69.

Rajaram, P. K., and C. Grundy-Warr, eds. 2007. *Borderscapes: Hidden Geographies and Politics at Territory's Edge*. Minneapolis: University of Minnesota Press.

Ramadan, Tariq. 2012. *The Arab Awakening: Islam and the New Middle East*. New York: Allen Lane.

Rancière, Jacques. 1995. *On the Shores of Politics*. New York: Verso.

———. 2001. 'Ten Theses on Politics'. *Theory and Event* 5(3). Accessed 15 January 2014. www.ucd.ie/philosophy/staff/maevecooke/Ranciere.Ten.pdf

———. 2004a. *Disagreement: Politics and Philosophy*. Minneapolis: University of Minnesota Press.

———. 2004b. *The Politics of Aesthetics: The Distribution of the Sensible*. London: Continuum.

Read, Jason. 2003. *The Micro-Politics of Capital: Marx and the Prehistory of the Present*. Albany: State University of New York Press.

———. 2009. 'A Genealogy of Homo-Economicus: Neoliberalism and the Production of Subjectivity'. *Foucault Studies* 6: 25–36.

Revel, Judith. 2008a. Resistances, Subjectivities, Common. *Generation Online*. Accessed 19 March 2014. http://www.generation-online.org/p/fprevel4.htm

———. 2008b. 'Identità, natura, vita. Tre decostruzioni biopolitiche'. In *Foucault oggi*, ed. M. Galzigna. Milano: Feltrinelli.

———. 2010. *Foucault, une pensée du discontinu*. Paris: Mille et une nuits.

———. 2011. Risposte al forum 'Michel Foucault e le resistenze'. In *Materialifoucaultiani*. Accessed 3 March 2014. http://www.materialifoucaultiani.org/it/materiali/altri-materiali/59-forum-qmichel-foucault-e-le-resistenzeq/153-materiali-foucaultiani-judith-revel-1.html

———. 2012. 'Diagnosi, soggettivazione, comune: Tre faccie dell'emancipazione oggi'. Uninomade. Accessed 19 March 2014. http://uninomade.org/wp/wp-content/uploads/2012/05/Diagnosi-soggettivazione-comune-tre-faccie-dellemncipazione-oggi.pdf

———. 2013a. 'What Are We at the Present Time? Foucault and the Question of the Present'. Paper presented at Goldsmiths College, workshop on Foucault and the History of the Present, 28 February.

———. 2013b. 'La violence et ses formes'. *Rue Descartes* 77(1). Accessed 18 March 2014. http://nomodos.blogspot.it/2013/05/rue-descartes-n77-20131-pouvoir.html

———. 2013c. 'Foucault and His 'Other': Subjectivation and Displacement'. In *The Biopolitics of Development: Reading Foucault in the Postcolonial Present*, edited by Sandro Mezzadra, Julian Reid and Ranabbir Samaddar, 15–24. New Delhi: Springer.

Rigo, Enrica. 2008. *Europa di confine: Trasformazioni della cittadinanza nell'Unione allargata*. Roma: Meltemi.

Rodier, Claire. 2006. 'Analysis of the External Dimension of the EU's Asylum and Immigration Policies'. Summary and recommendations for the European Parliament. Accessed 15 March 2014. www.europarl.europa.eu/meetdocs/2004_2009/documents/dt/619/619330/619330en.pdf

———. 2012. *Xénophobie business: À quoi servent les contrôles migratoires?* Paris: Editions La Decouverte.

Rodriguez, Nunes. 1996. 'The Battle for the Border: Notes on Autonomous Migration, Transnational Communities, and the State'. *Social justice* 23(3): 21–37.

Rose, Nikolas. 1996a. *Inventing Our Selves: Psychology, Power, and Personhood*. Cambridge: Cambridge University Press.

———. 1996b. Governing 'Advanced' Liberal Democracies, in *Foucault and Political Reason: Liberalism, Neo-Liberalism and Rationalities of Government*, 37–64. Chicago: University of Chicago Press.

———. 1999. *Powers of Freedom: Reframing Political Though*t. Cambridge: Cambridge University Press.

Rose, Nikolas, Pat O'Malley and Mariana Valverde. 2006. 'Governmentality'. *Annual Review of Law and Social Science* 2: 83–104.

Rudnyckyj, Daromir. 2004. 'Technologies of Servitude: Governmentality and Indonesian Transnational Labor Migration'. *Anthropological Quarterly* 77(3): 407–34.

Rumford, Christian. 2006. 'Theorizing Borders'. *European Journal of Social Theory* 9(2): 155–69.

Rygel, Kim. 2011. *Globalizing Citizenship*. Vancouver: University of British Columbia Press.

Sakai, Naoki, and Jon Solomon, eds. 2006. *Translation, Biopolitics, Colonial Difference*. Hong Kong: Hong Kong University Press.
Samaddar, Ranabbir. 2009. *Emergence of the Political Subject*. New Delhi: Sage.
Samers, Michael. 2004. 'An Emerging Geopolitics of 'Illegal' Immigration in the European Union'. *European Journal of Migration and Law* 6(1): 27–45.
Sanyal, Kanyal. 2007. *Rethinking Capitalist Development: Primitive Accumulation, Governmentality and the Post-Colonial Capitalism*. London: Routledge.
Sassen, Saskia. 2006. *Territory, Authority, Rights: From Medieval to Global Assemblages*. Princeton, NJ: Princeton University Press.
———. 2013. 'When Territory Deborders Territoriality'. *Territory, Politics, Governance* 1(1): 21–45.
Sayad, Abdelmalek. 2004. *The Suffering of the Immigrant*. Cambridge: Polity.
Schuster, Liza. 2011. 'Dublin II and Eurodac: Examining the (un)intended(?) Consequences'. *Gender, Place & Culture: A Journal of Feminist Geography*18(3): 401–16.
Sciurba, Alessandra. 2009. *Campi di forza: Percorsi confinati di migranti in Europa*. Verona: Ombre Corte.
Selby, Jan. 2007. 'Engaging Foucault: Discourse, Liberal Governance and the Limits of Foucauldian IR'. *International Relations* 21(3): 324–45.
Selmeczi, Anna. 2009. '". . . we are being left to burn because we do not count": Biopolitics, Abandonment, and Resistance'. *Global Society* 23(4): 519–38.
Seth, Sanjay. 2007. *Subject Lessons: The Western Education of Colonial India*. Durham, NC: Duke University Press.
Shukaitis, Stevphen, David Graeber, and Erika Biddle, eds. 2007. *Constituent Imagination: Militant Investigations, Collective Theorization*. Oakland, CA: AK Press.
Sidaways, Jon. 2007. 'Spaces of Postdevelopment'. *Progress in Human Geography* 31(3): 345–61.
Smith, David. 2000. *Moral Geographies: Ethics in a World of Difference*. Edinburgh: Edinburgh University Press.
Soguk, Nevzat. 2007. 'Border's Capture: Insurrectional Politics, Border-Crossing Humans, and the New Political'. In *Borderscapes: Hidden Geographies and Politics at Territory's Edge*, edited by Prem Kumar Rajaram and Carl Grundy-Warr, 283–308. Minneapolis: University of Minnesota Press.
Sorgoni, Barbara. 2011. 'Pratiche ordinarie per presenze straordinarie. Accoglienza, controllo e soggettività nei Centri per richiedenti asilo in Europa'. *Lares* 1: 19–38.
Sossi, Federica. 2002. *Autobiografie negate: Immigrati nei lager del presente*. Roma, Manifestolibri.
———. 2007. *Migrare: Spazi di confinamento e strategie di esistenza*. Milano: IlSaggiatore.
———. 2012. 'Qui e lì sono la Stessa Cosa. Sommovimenti di spazi e narrazioni'. In *Spazi in migrazione: Cartoline di una rivoluzione*, edited by Sossi Federica. Verona: Ombre Corte, Verona.
———. 2013a. 'Struggles in Migration: The Phantoms of Truths'. In *Spaces in Migration: Postcards of a Revolution*, edited by Glenda Garelli, Federica Sossi and Martina Tazzioli. London: Pavement Books.
———. 2013b. 'Cartographic Games in-between an Upheaval's Folds'. In *Spaces in Migration: Postcard of a Revolution*, edited by Glenda Garelli, Federica Sossi and Martina Tazzioli. London: Pavement Books.
Sparke, Matthew. 2006. 'A Neoliberal Nexus: Citizenship, Security and the Future of the Border.' *Political Geography* 25(2): 151–80.
Spijkerboer, Thomas. 2007. 'The Human Costs of Border Control'. *European Journal of Migration and Law* 9: 127–39.
Spivak, Gayatri C. 1999. *A Critique of Postcolonial Reason: Toward a History of the Vanishing Present*. Cambridge, MA: Harvard University Press.
Squire, Vicki. 2009. *The Exclusionary Politics of Asylum*, Basingstoke: Palgrave.
Squire, Vicki, ed. 2011. *The Contested Politics of Mobility: Borderzones and Irregularity*. Abingdon, UK: Routledge.

Stephen, Lynn. 2008. 'Los nuevos desaparecidos y muertos: Immigration, Militarization, Deaths and Disappearances on Mexico's Border'. Texas Civil Rights Project. Accessed 25 March 2014. http://texascivilrightsproject.org/docs/hr/tcrp2013migrant.pdf

Stoler, Ann L. 1996. *Race and the Education of Desire: Foucault's History of Sexuality and the Colonial Order of Things*. Durham, NC: Duke University Press.

———. 2002. *Carnal Knowledge and Imperial Power: Race and the Intimate in Colonial Rule*. Berkeley: University of California Press.

Tapinos, George P. 1990. 'Development Assistance Strategies and Emigration Pressure in Europe and Africa'. Commission for the Study of International Migration and Co-operative Economic Development.

Tazzioli, Martina. 2012. 'Choucha, Ras-Jadir, Tatouine: Sconfinamenti degli spazi-frontiera'. In *Spazi in migrazione: Cartoline di una rivoluzione*, edited by Federica Sossi. Verona: Ombre Corte.

Tondini Matteo. 2010. 'Fishers of Men? The Interception of Migrants in the Mediterranean Sea and Their Forced Return to Libya'. INEX paper. Social Science Research Network. Accessed 20 March 2014. http://ssrn.com/abstract=1873544

Truong, Tahnh-Dam, and Des Gasper, eds. 2011. *Transnational Migration and Human Security: The Migration-Development-Security Nexus*. Heidelberg: Springer.

Tsianos, Vassilis. 2007. 'Imperceptible Politics: Rethinking Radical Politics of Migration and Precarity Today'. PhD dissertation, University of Hamburg.

Tsing, Anna. 1997. 'The Global Situation'. *Cultural Anthropology* 15(3): 327–60.

Turnbull, David. 2000. *Masons, Tricksters and Cartographers: Comparative Studies in the Sociology of Scientific and Indigenous Knowledge*. London: Taylor & Francis.

Van der Ploeg, Irma. 1999. 'The Illegal Body: "Eurodac" and the Politics of Biometric Identification'. *Ethics and Information Technology* 1(4): 295–302.

Van Houtum, Henk. 2002. 'Bordering, Ordering and Othering'. *Tijdschrift voor economische en sociale geografie* 93(2): 125–36.

———. 2005. 'The Geopolitics of Borders and Boundaries'. *Geopolitics* 10: 672–67.

Van Munster, Rens. 2009. *Securitizing Immigration: The Politics of Risk in the EU*. London: Palgrave Macmillan.

Vaughan, Megan. 1991. *Curing Their Ills: Colonial Power and African Illness*. Stanford, CA: Stanford University Press.

Vermeren, Pierre. 2011. *Maghreb : Les origines de la révolution démocratique*. Paris: Pluriel.

Virno, Paolo. 2002. *Grammatica della moltitudine: Per un'analisi delle forme di vita contemporanee*. Roma: DeriveApprodi.

Vrasti, Wanda. 2013. 'Universal but Not Truly 'Global': Governmentality, Economic Liberalism, and the International'. *Review of International Studies* 39(1): 49–69.

Walker, Robert B. J. 2000. 'Europe Is Not Where It Is Supposed to Be'. In *International Relations Theory and the Politics of European Integration: Power, Security and Community*, edited by Morten Kelstrup and Michael Williams, 14–32. London: Routledge.

Walters, William. 2002a. 'Mapping Schengenland: Denaturalizing the Border'. *Environment and Planning D: Society and Space* 20(5): 561–80.

———. 2002b. 'Deportation, Expulsion, and the International Police of Aliens'. *Citizenship Studies* 6(3): 265–91.

———. 2006. 'Border/Control'. *European Journal of Social Theory* 9(2): 187–204.

———. 2008. 'Bordering the Sea: Shipping Industries and the Policing of Stowaways'. *Borderlands E-journal* 7(3). Accessed 20 March 2014. http://www.borderlands.net.au/vol7no3_2008/walters_bordering.htm

———. 2009. 'Anti-Political, Economy: Cartographies of 'Illegal Immigration' and the Displacement of the Economy'. In *Cultural Political Economy*, edited by Jacquelin Best and Matthew Peterson, 113–38. London: Routledge.

———. 2011a. 'Foucault and Frontiers: Notes on the Birth of the Humanitarian Border'. In *Governmentality: Current Issues and Future Challenges*, edited by Ulrich Brockling, Susanne Krasmann and Thomas Lemke. New York: Routledge, 138–64.

———. 2011b. 'Rezoning the Global: Technological Zones, Technological Work and the (Un)Making of Biometric Borders'. In *The Contested Politics of Mobility: Borderzones and Irregularity*, edited by Vicki Squire, 51–73. London: Routledge.
———. 2012. *Governmentality: Critical Encounters*. Oxford: Routledge.
———. 2013. 'Forum on Foucault, Migrations and Borders'. *Materialifoucaultiani* 3. Accessed 18 March 2014. http://www.materialifoucaultiani.org/en/rivista/volume-ii-numero-3.html
Wahnich, Sophie. 1997. *L'impossible citoyen: L'étranger dans le discours de la révolution française*. Paris: Albin Michel.
Watson, Scott. 2013. *The Securitization of Humanitarian Migration: Digging Moats and Sinking Boats*. London: Routledge.
Watts, Michael. 2003. 'Development and Governmentality'. *Singapore Journal of Tropical Geography* 24(1): 49–60.
Weinzierl, Ruth, and Urszula Lisson. 2007. *Border Management and Human Rights: A Study of EU Law and the Law of the Sea*. German Institute for Human Rights. Accessed 15 March 15 2014. www.institut-fuer-menschenrechte.de
Winichakul, Thongchai. 1997. *Siam Mapped: A History of the Geo-Body of a Nation*. Honolulu: University of Hawaii Press.
Wolff, Sarah. 2008. 'Border Management in the Mediterranean: Internal, External and Ethical Challenges'. *Cambridge Review of International Affairs* 21(2): 253–70.
Wood, Denis. 1992. *The Power of Maps*. New York: Guilford Press.
Young, Robert. 2001. *Postcolonialism: An Historical Introduction*. Oxford: Blackwell.
Zetter, Robert. 2007. 'More Labels, Fewer Refugees: Remaking the Refugee Label in an Era of Globalization'. *Journal of Refugee Studies* 20(2): 172–92.
Žižek, Slavoi. 2005. 'Against Human Rights'. *New Left Review* 34: 131–35.

INSTITUTIONAL DOCUMENTS

(EC) No. 2725/2000
COM (2005) 388 final
Council of Europe (1471) 2005
COM (2009) 447 final
COM (2011) 200 final
COM (2011) 560 final
Spring Program, MEMO/11/636, 2011
MEMO/11/642, 2011
COM (2011) 873 final
COM (2011) 248 final
EU-IOM, December 2011
Parliamentary Assembly, Doc. 12628/2011
EU-Lisa (2013)
COM (2012) 533 final
COM (2013) 197 final, 2013/0106(COD)
Easo-Frontex (2012) Working Arrangement
FRA (European Union Agency for Fundamental Rights) 2013
IOM (2011) START
IOM (2012a) Migrants Caught in Crisis: The IOM Experience in Libya

IOM (2012b) Protecting Migrants during Times of Crisis: Immediate Responses and Sustainable Strategies, International Dialogue on Migration 2012 Managing Migration in Crisis Situations

IOM (2012c) Moving to Safety: Migration Consequences of Complex Crisis, International Dialogue on Migration 2012 Managing Migration in Crisis Situations

IOM (2012d) IOM Migration Crisis Operational Framework

IMO (1974/1980) International Convention for the Safety of Life at Sea (SOLAS)

United Nations Convention on the Law of the Sea of 10 December 1982

UNHCR (1991) Background Note on the Safe Country Concept and Refugee Status, 26 July 1991, EC/SCP/68

UNHCR (2007) Response to the European Commission's Green Paper on the Future Common European Asylum System

UNHCR (2007) Advisory Opinion on the Extraterritorial Application of Non-Refoulement Obligations under the 1951 Convention Relating to the Status of Refugees and Its 1967 Protocol

Index